SpringerBriefs in Mathematical Physics

Volume 41

SpringerBriefs are characterized in general by their size (50–125 pages) and fast production time (2–3 months compared to 6 months for a monograph).

Briefs are available in print but are intended as a primarily electronic publication to be included in Springer's e-book package.

Typical works might include:

- An extended survey of a field
- A link between new research papers published in journal articles
- A presentation of core concepts that doctoral students must understand in order to make independent contributions
- Lecture notes making a specialist topic accessible for non-specialist readers.

SpringerBriefs in Mathematical Physics showcase, in a compact format, topics of current relevance in the field of mathematical physics. Published titles will encompass all areas of theoretical and mathematical physics. This series is intended for mathematicians, physicists, and other scientists, as well as doctoral students in related areas.

Editorial Board

- Nathanaël Berestycki (University of Cambridge, UK)
- Mihalis Dafermos (University of Cambridge, UK / Princeton University, US)
- Atsuo Kuniba (University of Tokyo, Japan)
- Matilde Marcolli (CALTECH, US)
- Bruno Nachtergaele (UC Davis, US)
- Hal Tasaki (Gakushuin University, Japan)

- 50 – 125 published pages, including all tables, figures, and references
- Softcover binding
- Copyright to remain in author's name
- Versions in print, eBook, and MyCopy

More information about this series at http://www.springer.com/series/11953

Blake C. Stacey

A First Course
in the Sporadic SICs

 Springer

Blake C. Stacey
University of Massachusetts Boston
Boston, MA, USA

ISSN 2197-1757 ISSN 2197-1765 (electronic)
SpringerBriefs in Mathematical Physics
ISBN 978-3-030-76103-5 ISBN 978-3-030-76104-2 (eBook)
https://doi.org/10.1007/978-3-030-76104-2

This Springer imprint is published by the registered company Springer Nature Switzerland AG
The registered company address is: Gewerbestrasse 11, 6330 Cham, Switzerland

Preface

Physics has social and aesthetic aspects that many of its most vocal boosters are loath to admit. Even when we physicists let down our hair in memoirs and interviews and other semi-informal venues and admit that the beauty of our subject moves us, we are reluctant to confess that the "beauty" we invoke is an intensely personal quality. A matter of taste, to be frank about it. To me, SICs offer the enigmatic appeal of patterns halfway seen, like a forest in a predawn fog.

This book is not about completed patterns. We do not know how many SICs there are, and even if we narrow our attention to the sporadic SICs, the primary focus of these chapters, there are plenty of unresolved questions on the edges of what we will cover here. It is even possible that the classification of sporadic SICs is incomplete, and a structure found tomorrow will have to be added to the list. I have tried to point out these questions where I could. Many of the exercises are not more than a step or two from the research frontier.

It is my hope that you have at least one reaction to this book that I did not intend. A certain amount of vagueness is necessary, I think sometimes, for fruitful scientific communication. For if my statement creates precisely the consequences I planned, matching my intention in every detail, then there would be no room for spontaneity, for the generation of leaps I myself was unable to make. There is no point in your thinking *only* what I have already thought! The challenge lies in making those moments of imprecise fantasy, of ideas-for-an-idea, fall in the appropriate places amid the direct and declarative surroundings. "To construct a pleasingly thrilling contrapuntal structure," to sound like a person who thinks they know music theory.

The conceptual background to my work with SICs is the research program known as QBism, an ongoing project in quantum foundations whose goal is to identify the characteristics of nature which make quantum theory such a good tool for navigating in it. For the most part, this book will skim over the surface of those philosophical waters, as they run quite deep and my page count here is quite finite. I have also mostly elided the topic of how to implement in the laboratory the measurements defined by SICs. This is for the very good reason that I am a theorist and should be kept as far away from laboratories as possible. I am not the best one to explain how to build machines that go *click*, but I know that experimentalists are clever folk.

To suggest that having four different kinds of click instead of two is an impossible extravagance would insult their capabilities.

I would like to thank my family for keeping in touch at a safe distance, and my housemates for putting up with me in close proximity. My collaborators deserve more gratitude than a preface can contain. I hesitate to list names in the certainty that I will omit an important one, but I know I cannot let this moment pass without expressing my appreciation for Marcus Appleby, Gabriela Barreto Lemos, John B. DeBrota, Christopher A. Fuchs, Jacques L. Pienaar, and Huangjun Zhu. Back in the days when we had conferences, I had wonderful learning experiences about SICs at Worcester Polytechnic Institute, the Ohio State University, two March Meetings of the American Physical Society, and at MIT. More recently, we have had video chats, upon occasion managing to bridge four continents. As these meetings have been a highlight of the past year, I owe a kind word to the participants, including Ingemar Bengtsson, Irina Dumitru, Mary Fries, Amanda Gefter, Sachin Gupta, Bob Henderson, Kim Reece, Kathryn Schaffer, Juan Varela, and Matthew Weiss. John Baez and Karol Życzkowski gave me opportunities to expound on these ideas in blog posts and video conference, respectively. Some early drafts of what would become this book were written while I had the support of the John Templeton Foundation; all of the opinions expressed here are my own and not those of the JTF. David Harden and Eric Downes saw this book in manuscript and provided corrections. If any errors remain, Eric knows that I will transfer all the responsibilities to him; we met in college, and he is well aware that I have no shame.

I found my way to SICs many years ago, during a time when life had left me emotionally adrift. They turned out to be the research problem I needed. I do not know whether they can provide anyone else a moment of respite, but this book is a good opportunity to find out.

Near Boston, MA, USA   Blake C. Stacey

Contents

1 Equiangular Lines 1
   1.1 Introduction 1
   1.2 Real Lines 2
   1.3 Complex Lines 6
   References 10

2 Optimal Quantum Measurements 13
   2.1 Introduction 13
   2.2 SIC Representations of Quantum States 15
   2.3 Constructing SICs Using Groups 22
   References 24

3 Geometry and Information Theory for Qubits and Qutrits 27
   3.1 Qubits 27
   3.2 Qutrits 28
   3.3 Coherence 31
   References 36

4 SICs and Bell Inequalities 39
   4.1 Mermin's Three-Qubit Bell Inequality 40
   4.2 The Hoggar SIC 41
   4.3 Qubit Pairs and Twinned Tetrahedral SICs 45
   4.4 Failure of Hidden Variables for Qutrits 50
   4.5 Quantum Theory from Nonclassical Probability Meshing 52
   References 53

5 The Hoggar-Type SICs 57
   5.1 Introduction 57
   5.2 Simplifying the QBic Equation 59
   5.3 Triple Products and Combinatorial Designs 60
   5.4 The Twin of the Hoggar SIC 67
   5.5 Combinatorial Designs from the Twin Hoggar SIC 69
   5.6 Quantum-State Compatibility 72
   5.7 From Pauli Operators to Real Equiangular Lines 78

   5.8   Concluding Remarks 80
   References 81

6 Sporadic SICs and the Exceptional Lie Algebras 83
   6.1   Root Systems and Lie Algebras 83
   6.2 E_6 86
   6.3 E_8 88
   6.4 E_7 90
   6.5   The Regular Icosahedron and Real-Vector-Space Quantum
        Theory 92
   6.6   Open Puzzles Concerning Exceptional Objects 95
   References 100

7 Exercises 103
   References 111

Index 113

Chapter 1
Equiangular Lines

I introduce the problem of finding maximal sets of equiangular lines, in both its real and complex versions, attempting to write the treatment that I would have wanted when I first encountered the subject. Equiangular lines intersect in the overlap region of quantum information theory, the octonions and Hilbert's twelfth problem. The question of how many equiangular lines can fit into a space of a given dimension is easy to pose—a high-school student can grasp it—yet it is hard to answer, being as yet unresolved. This contrast of ease and difficulty gives the problem a classic charm.

1.1 Introduction

To motivate the definition, we can start with the most elementary example: the diagonals of a regular hexagon. Any two of them cross and create what the schoolbooks call supplementary vertical angles. Without loss of information, we can take "the" angle defined by the pair of lines to be the smaller of these two values. Moreover, this value is the same for all three possible pairs of lines: For any two diagonals, their angle of intersection will be $\pi/3$. We can state this in a way amenable to generalization if we lay a unit vector along each of the three diagonals. Whichever way we choose to orient the vectors, their inner products will satisfy

$$|\langle \mathbf{v}_j, \mathbf{v}_k \rangle| = \begin{cases} 1, & j = k; \\ \alpha, & j \neq k. \end{cases} \qquad (1.1)$$

When a set of vectors $\{\mathbf{v}_j : j = 1\ldots, N\}$ enjoys this property, it yields a set of *equiangular lines*. An orthonormal basis is equiangular, with $\alpha = 0$. The question becomes more intriguing when we push the size N of the set beyond the dimension d. For example, if we step from $\mathbb{R}^2$ up to $\mathbb{R}^3$, it is already hard to guess how many

© The Author(s), under exclusive license to Springer Nature Switzerland AG 2021
B. C. Stacey, *A First Course in the Sporadic SICs*,
SpringerBriefs in Mathematical Physics 41,
https://doi.org/10.1007/978-3-030-76104-2_1

equiangular lines we can pack in, and what that configuration might look like. A decent first move would be to consider the Platonic solids, since they are nicely symmetrical. What would make Kepler happy? It turns out that this is not a bad strategy to pursue: The diagonals of a regular icosahedron are a set of six equiangular lines in $\mathbb{R}^3$. This allows us to express our lines using the golden ratio, so if we are unscrupulous we can surely make a buck off somebody. More significantly, both the hexagon in $\mathbb{R}^2$ and the icosahedron in $\mathbb{R}^3$ give sets that are *maximal*. There is no way to get more than $N = 3$ equiangular lines in two dimensions, or more than $N = 6$ in three dimensions.

We can now mathematician up the question in two ways. First, we can generalize to arbitrary dimension and ask how the maximum N varies with d. Second, we can replace the real numbers $\mathbb{R}$ with the complex numbers $\mathbb{C}$, because our definition of equiangularity (1.1) works just as well with both. It is the latter move which makes this topic of geometry also a subject for the quantum physicist.

One can without too much pain prove the *Gerzon bound*: The size of a set of equiangular lines cannot exceed $d(d + 1)/2$ in $\mathbb{R}^d$ or d^2 in $\mathbb{C}^d$ [1]. When the Gerzon bound is met, the value of α is fixed, to $1/\sqrt{d + 2}$ in $\mathbb{R}^d$ and $1/\sqrt{d + 1}$ in $\mathbb{C}^d$. In the real case, we know that we cannot in general attain the Gerzon bound, and the question of how big N can be as a function of d ties in with some of the most remarkable structures in discrete mathematics. Meanwhile, in the complex case, it *appears* that we can attain the Gerzon bound in every dimension, but decades of work have not yet settled the matter one way or the other—and what we have seen so far has already led us into deep questions of number theory and quantum mechanics.

1.2   Real Lines

We take up the real case first. In general, the discrete choice of sign factors made when picking unit vectors to represent lines gives the study of maximal equiangular line-sets in $\mathbb{R}^d$ a very combinatorial flavor, and the theory of finite simple groups plays an intriguing role. A mere sampling of the known solutions and the topics related to them will, regrettably, have to suffice. The three diagonals of a regular hexagon are a maximal set of equiangular lines in $\mathbb{R}^2$, saturating the Gerzon bound. Likewise, the six diagonals of a regular icosahedron attain the Gerzon bound in $\mathbb{R}^3$. We start to fall short at dimension $d = 4$, where it turns out we cannot exceed $N = 6$. The only two other cases where the Gerzon bound can be reached, as far as anyone knows, are in $d = 7$ and $d = 23$. While the regular icosahedron dates back to the semi-legendary centuries of ancient mathematics, thinking in terms of maximal sets of equiangular lines is nowadays credited to Haantjes, who solved the $\mathbb{R}^2$ and $\mathbb{R}^3$ cases in 1948. Van Lint and Seidel resolved $\mathbb{R}^4$ through $\mathbb{R}^7$ in 1966 [2]. Even today, uncertainty starts to creep in as soon as dimension $d = 17$ [3]. The Gerzon bound can only be attained above $d = 3$ if $d + 2$ is the square of an odd integer, but not all odd integers qualify: The maximum is known to be strictly less than the Gerzon bound in dimensions 47 and 79 [4, 5].

But how remarkable those solutions in $d = 7$ and $d = 23$ are! These sets of equiangular lines can be extracted from celebrated structures in one dimension higher, the E_8 and Leech lattices respectively. The $N = 28$ lines in $\mathbb{R}^7$ are the diagonals of the *Gossett polytope* 3_{21}, and they correspond among other things to the 28 bitangents to a general plane quartic [6]. One way to obtain these lines—there are different constructions, but the results are all equivalent up to an overall rotation—stems from an observation by Van Lint and Seidel [2, 7] that if we want an interesting set that involves the number seven, we ought to turn to the Fano plane sooner or later. This geometry (the "combinatorialist's coat of arms") is a set of seven points grouped into seven lines such that each line contains three points and each point lies within three distinct lines, with each pair of lines intersecting at a single point. Consequently, if we take the *incidence matrix* of the Fano plane, writing a 1 in the ijth entry if line i contains point j, then every two rows of the matrix have exactly the same overlap:

$$M = \begin{pmatrix} 1 & 1 & 1 & 0 & 0 & 0 & 0 \\ 1 & 0 & 0 & 1 & 1 & 0 & 0 \\ 1 & 0 & 0 & 0 & 0 & 1 & 1 \\ 0 & 1 & 0 & 1 & 0 & 1 & 0 \\ 0 & 1 & 0 & 0 & 1 & 0 & 1 \\ 0 & 0 & 1 & 1 & 0 & 0 & 1 \\ 0 & 0 & 1 & 0 & 1 & 1 & 0 \end{pmatrix}. \tag{1.2}$$

The rows of the incidence matrix furnish us with seven equiangular lines in $\mathbb{R}^7$. To build this out into a full set of 28 lines, we can introduce sign factors [2, 7], and one way to do that is to add *orientations* to the Fano plane, exactly as one does when using it as a mnemonic for octonion multiplication. We can label the seven points with the imaginary octonions e_1 through e_7. When drawn on the page, a useful presentation of the Fano plane has the point e_4 in the middle and, reading clockwise, the points e_1, e_7, e_2, e_5, e_3 and e_6 around it in a regular triangle. (See Fig. 1.1.) The three sides and three altitudes of this triangle, along with the inscribed circle, provide the seven Fano lines: (e_1, e_2, e_3), (e_1, e_4, e_5), (e_1, e_7, e_6), (e_2, e_4, e_6), (e_2, e_5, e_7), (e_3, e_4, e_7), (e_3, e_6, e_5). The sign of a product depends upon the order, for example, $e_1 e_2 = e_3$ but $e_2 e_1 = -e_3$. The full multiplication table, due to Cayley and Graves, is

$e_i e_j$	1	e_1	e_2	e_3	e_4	e_5	e_6	e_7
1	1	e_1	e_2	e_3	e_4	e_5	e_6	e_7
e_1	e_1	-1	e_3	$-e_2$	e_5	$-e_4$	$-e_7$	e_6
e_2	e_2	$-e_3$	-1	e_1	e_6	e_7	$-e_4$	$-e_5$
e_3	e_3	e_2	$-e_1$	-1	e_7	$-e_6$	e_5	$-e_4$
e_4	e_4	$-e_5$	$-e_6$	$-e_7$	-1	e_1	e_2	e_3
e_5	e_5	e_4	$-e_7$	e_6	$-e_1$	-1	$-e_3$	e_2
e_6	e_6	e_7	e_4	$-e_5$	$-e_2$	e_3	-1	$-e_1$
e_7	e_7	$-e_6$	e_5	e_4	$-e_3$	$-e_2$	e_1	-1

$$\tag{1.3}$$

Fig. 1.1 The Fano plane,
with its points labeled e_1
through e_7

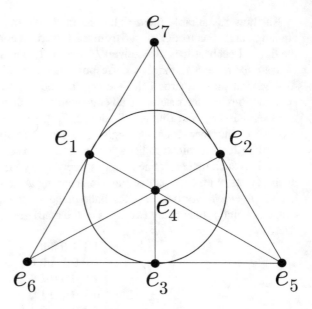

which we can express visually by carefully assigning arrows to the lines of the Fano
plane.

To build our set of 28 equiangular vectors, start by taking the first row of the
incidence matrix M, which corresponds to the line (e_1, e_2, e_3), and give it all possible
choices of sign by multiplying by the elements not on that line. Multiplying by e_4,
e_5, e_6 and e_7 respectively, we get

$$\begin{pmatrix} + & + & + & 0 & 0 & 0 & 0 \\ - & + & - & 0 & 0 & 0 & 0 \\ - & - & + & 0 & 0 & 0 & 0 \\ + & - & - & 0 & 0 & 0 & 0 \end{pmatrix}. \tag{1.4}$$

Doing this with all seven lines of the Fano plane, we obtain a set of 28 vectors, each
one given by a choice of a line and a point not on that line. For any two vectors
derived from the same Fano line, two of the terms in the inner product will cancel,
leaving an overlap of magnitude 1. And for any two vectors derived from different
Fano lines, the overlap always has magnitude 1 because any two lines always meet
at exactly one point (Fig 1.1).

As we go up from $\mathbb{R}^7$ to $\mathbb{R}^{23}$, the properties tap into more of the esoteric (Table 1.1).
As mentioned above, we can obtain the $N = 276$ equiangular lines living in $\mathbb{R}^{23}$ from
the Leech lattice. This lattice is how a grocer would stack 24-dimensional oranges;
the points of the lattice are the locations of the centers of the spheres in the densest
possible packing thereof in 24 dimensions [8]. The vectors from the origin to the
lattice points are classified by their "type", which is half their norm. To obtain a
Gerzon-bound-saturating set of equiangular lines, start with a vector of type 3. For

Table 1.1 Bounds on the largest possible size of a set of equiangular lines in $\mathbb{R}^d$. For some values of the dimension d, the bound is not known exactly. Dimensions where the Gerzon bound is saturated are shown in bold. For more details, see OEIS:A002853 and references therein

d	2	3	4	5	6	7–14	15	16	17
$N_{\max}(d)$	3	6	6	10	16	28	36	40	48–49
d	18	19	20	21	22	23–41	42	43	
$N_{\max}(d)$	56–60	72–75	90–95	126	176	276	276–288	344	

any such vector $\mathbf{v}$, there are 276 unordered pairs of other lattice vectors having minimal norm (type 2) that add to $\mathbf{v}$. Each pair specifies a line, and we obtain 276 lines in all. These lines have rich group-theoretic significance [1, 9]. First of all, their symmetry group is Conway's third sporadic group Co_3. The *stabilizer* of a line is the subgroup comprising those symmetries of the set that leave that line fixed while permuting the others. For the $N = 276$ lines in $\mathbb{R}^{23}$, the stabilizer of any line is isomorphic to the McLaughlin sporadic simple group McL. Moreover, if we find the largest possible subset of the $N = 276$ lines that are all orthogonal to a common vector in $\mathbb{R}^{23}$, we obtain a set of 176 vectors that all squeeze into $\mathbb{R}^{22}$, and these furnish a maximal set of equiangular lines in that dimension. The symmetries of this set constitute the Higman–Sims sporadic simple group HS. By close consideration of the ways in which vector directions can be assigned to the $N = 276$ lines, it is also possible to distinguish special subsets of 23 lines and thence obtain the Mathieu group M_{23} [10].

We do not yet have a general theory of how the maximum N varies with the dimension d, and of course, we can only hazard guesses about what the textbooks of tomorrow might contain, but one conceptual connection that has proved quite important so far is to algebraic graph theory. If $\{\mathbf{v}_i : i = 1, \ldots, N\}$ is a set of unit vectors defining a set of equiangular lines, then by the definition of equiangularity, the Gram matrix of these unit vectors will be 1 along the diagonal and $\pm\alpha$ everywhere else. Note that specifying a choice of sign for each off-diagonal entry is exactly the same information needed to specify which edges are connected in a graph with N vertices. Changing the sign of any vector will flip the signs on some of the entries in the Gram matrix, which rewires the corresponding graph in a particular way. So, a set of N equiangular lines corresponds to an equivalence class of N-vertex graphs (a "switching class"), and there is a natural overlap of interest between equiangular lines and strongly regular graphs [5, 10–12].

Because the matrices one tries to construct when attempting to build sets of real equiangular lines will be filled with integers, conditions for their existence will be equations that algebraic number theory is suited to handle. In particular, the study of cyclotomic fields—fields made by extending $\mathbb{Q}$ with a primitive nth root of unity—becomes relevant [12].

One fruitful avenue of inquiry has been to *fix an angle* and ask how many lines in $\mathbb{R}^d$ can be equiangular with that chosen angle [4, 5, 13]. Balla et al. have proved that for a fixed α, when the dimension d becomes sufficiently large there can be at most $2(d-1)$ lines in $\mathbb{R}^d$ that are equiangular with common overlap α [14].

For students who wish to jump in and do calculations as quickly as possible, an arXiv paper by Tremain [15] provides a useful collection of constructions.

In this book, we will mostly focus on complex equiangular lines. However, some knowledge of the $\mathbb{R}^d$ version of the problem will turn out to be useful, because complex and real equiangular lines can relate in unexpected ways. In addition, bounds on N as a function of d are physically meaningful. Quantum theory employs complex vector spaces, as we will detail soon. One can devise, as a character foil to quantum physics, a theory constructed in the same manner but over real vector spaces instead. The fact that the Gerzon bound is attained in the complex case but not the real for some dimensions—possibly an infinite number of them—implies that measurements are possible in quantum theory which have no counterpart in the foil theory.

1.3   Complex Lines

The complex case was first studied as a natural counterpart to the real one. Investigations of structures like complex polytopes [16] turned up maximal sets of complex equiangular lines in dimensions $d = 2$, 3 and 8. Then the 1999 PhD thesis of Zauner [17], followed by the independent work of Renes et al. [18], made complex equiangular lines into a physics problem. Now we have exact solutions for 115 different values of the dimension, including all dimensions from 2 through 53 inclusive, and some as large as $d = 5799$. Furthermore, numerical solutions are known to high precision for all dimensions $d \leq 193$, and some as large as $d = 2208$ [19, 20]. These lists have grown irregularly, since different simplifications have proved applicable in different dimensions. (We endeavor to keep the website [21] up to date.) Credit is due to M. Appleby, I. Bengtsson, T.-Y. Chien, S. T. Flammia, M. Grassl, G. S. Kopp, A. J. Scott and S. Waldron.

Physicists know sets of d^2 equiangular lines in $\mathbb{C}^d$ as "Symmetric Informationally Complete Positive-Operator-Valued Measures", a mouthful that is abbreviated to *SIC-POVM* and often further just to *SIC* (pronounced "seek"). The "Symmetric" refers to the equiangularity property, while the rest summarizes the role these structures play in quantum theory [22, 23]. A basic premise of quantum physics is that to each physical system of interest is associated a complex Hilbert space $\mathcal{H}$. The subdisciplines of quantum information and computation [24] often employ finite-dimensional Hilbert spaces, $\mathcal{H}_d \simeq \mathbb{C}^d$. The mathematical representation of a *measurement process* is a set of positive semidefinite operators $\{E_j\}$ on $\mathcal{H}_d$ that sum to the identity. Each operator in the set stands for a possible outcome of the measurement, and the set as a whole is known as a POVM. To represent the preparation of a quantum system, we ascribe to the system a *density operator* ρ that is also a positive semidefinite operator on $\mathcal{H}_d$, in this case normalized so that its trace is unity. The *Born rule* states that the probability for obtaining an outcome of a measurement is given by the Hilbert–Schmidt inner product of the density operator and the POVM element representing that outcome:

$$p(j) = \text{tr}\, \rho E_j. \tag{1.5}$$

If the POVM $\{E_j\}$ has at least d^2 elements, then it is possible for it to span the space of Hermitian operators on $\mathcal{H}_d$. In this case, the POVM is *informationally complete* (IC), because any density operator can be expressed as its inner products with the POVM elements. Or, in more physical terms, the probabilities for the possible outcomes of an IC POVM completely specify the preparation of the quantum system. Given a set of unit vectors in $\mathbb{C}^d$ defining d^2 equiangular lines, the projectors $\{\Pi_j\}$ onto these lines span the Hermitian operator space, and up to normalization they provide a resolution of the identity:

$$\sum_{j=1}^{d^2} \Pi_j = dI. \tag{1.6}$$

SICs satisfy a host of optimality conditions. By many standards, a SIC is as close as one can possibly get to having an orthonormal basis for Hermitian operator space while staying within the positive semidefinite cone. This is highly significant for quantum theory, because positivity is a crucial aspect of having operator manipulations yield well-defined probabilities.

Decades of work on the fundamentals of quantum physics have shown that quantum uncertainty cannot be explained away as ignorance about "hidden variables" intrinsic to physical systems but concealed from our view. Historically, this area has been home to deep theorems like the results of Bell, Kochen and Specker [25], while in the modern age, it is also a topic of increasingly practical relevance, since if we want our quantum computers to be worth the expense, we had better understand exactly which phenomena can be imitated classically and which cannot [26]. SICs provide a new window on these questions by furnishing a measure of the margin by which any attempt to model quantum phenomena with intrinsic hidden variables is guaranteed to fail [27].

Here we have a peculiar confluence of topics advanced by the late John Conway. It was his insight into the Leech lattice that gave us the maximal equiangular set in $\mathbb{R}^{23}$, and together with Simon Kochen he carried forward the study of how hidden-variable hypotheses break down [28]. Somewhere in the world's weight of loss is the fact that we will never know what he might have thought about these two problems coming together.

Above, we noted the group-theoretic properties of real equiangular lines. Group theory also manifests in the complex case. First, all known SICs are *group covariant*, meaning that they can be generated by taking a well-chosen initial vector and computing its orbit under a group action. Moreover, in all cases except a class of solutions in $d = 8$, the group in question is the *Weyl–Heisenberg group* for dimension d. Given an orthonormal basis $\{e_n : i = 1, \ldots, d\}$, we define a shift operator X that sends e_n to e_{n+1} modulo d, and a "clock" or "phase" operator Z that sends e_n to $\exp(2\pi i n/d)e_n$. The two unitary operators X and Z commute up to the phase factor $\exp(2\pi i/d)$, and together with phase factors their products define the Weyl–Heisenberg group. (The $d = 2$ special case is also known as the Pauli group. Weyl introduced X and Z in 1925, in order to define what the quantum mechanics of a discrete degree of freedom could mean [29]. The association of Heisenberg's name with this group is a convention that has less historical justification, since the "canonical commutation relation" of position and momentum that inspired Weyl was not due to Heisenberg [30, 31].) The Weyl–Heisenberg group is significant in multiple aspects of quantum theory, such as the study of when quantum computations can be efficiently emulated classically. Zhu has proved that when the dimension d is prime, group covariance implies Weyl–Heisenberg covariance [32]. However, it is not known whether SICs must be group covariant in general.

In dimension $d = 8$, there exist Weyl–Heisenberg SICs but also a class of SICs covariant under a different group (the tensor product of three copies of the Pauli group). These were first discovered by Stuart Hoggar and can be termed the *Hoggar-type* SICs [33–35]. All of them are equivalent to one another under unitary and anti-unitary conjugations. Huangjun Zhu discovered that the stabilizer of an element in a Hoggar-type SIC is always isomorphic to PSU(3, 3), the group of 3×3 unitary matrices over the finite field of order 9 [36]. This is moreover the commutator subgroup of the automorphism group of the Cayley integral octonions [37], also known as the *octavians* [38]. In other words, the linear maps from the octavians to themselves that preserve the multiplication structure form a group, and an index-2 subgroup of that gives the ways to hold one vector in a Hoggar-type SIC in place and permute the other 63.

The octavians form a lattice, and up to an overall scaling, it is the same as the E_8 lattice. So, a symmetric arrangement of *complex lines* is, under the surface, tied in with an optimal packing of *real hyperspheres* [39]—a development that was completely unforeseen.

Exploration of the Weyl–Heisenberg SICs leads into algebraic number theory, and at a more subtle and demanding level than that subject has so far manifested in the study of real equiangular lines. During the early years, SIC vectors were found by computer algebra, laborious calculations with Gröbner bases yielding coefficients that were, in technical terms, ghastly [40]. Pages upon pages of printouts were sacrificed. Yet, upon closer study, these solutions turned out to be not as bad as they seemingly could have been: In $d \geq 4$, they were always expressible in terms of nested radicals. Since the polynomial equations being solved were of degrees much higher than quintic, there was no reason to expect a solution by radicals, and so the Galois theory of the SIC problem became a topic of interest [41]. This has led to a series of surprises.

Recall that a group G is solvable if we can write a series starting with the trivial group, $\langle 1 \rangle < H_1 < \cdots < H_m < G$, where each of the H_i is a normal subgroup of the next and the quotients H_i/H_{i-1} are all abelian. The zeros of a polynomial $f(x)$ over some field $\mathbb{K}$ can be found by radicals exactly when the splitting field of f, the extension of $\mathbb{K}$ in which f falls apart into linear factors, can be made by stacking up abelian extensions of $\mathbb{K}$. So, if we want to understand solvability by radicals, getting a handle on abelian extensions of number fields is the thing to do. One obvious place to start is abelian extensions of the rationals $\mathbb{Q}$, and the *Kronecker–Weber theorem* tells us that every abelian extension of $\mathbb{Q}$ is contained in a cyclotomic field.

Hilbert's twelfth problem asks for a broadening of the Kronecker–Weber theorem, or in other words, a classification of the abelian extensions of arbitrary number fields. This problem remains unresolved, although progress has been made. When we generalize beyond the case where the base field is $\mathbb{Q}$, the role of the cyclotomic fields is played by the *ray class fields*. The generalization of the n in a cyclotomic field's nth root of unity is a number called the *conductor* of the ray class field. In the original case covered by Kronecker–Weber, the fields in which the abelian extensions all live are generated by special values of a special function, i.e., the exponential function evaluated at certain points along the imaginary axis. What functions play the role of the exponential more generally, and at what points should they be evaluated? This is much more difficult to say.

Historically, the first to be understood beyond abelian extensions of the rationals themselves were abelian extensions of imaginary quadratic fields, that is, $\mathbb{Q}(\sqrt{-n})$ where n is a positive integer. This is significantly more demanding than the case where the base field is $\mathbb{Q}$. The theory that does the job goes by the name *complex multiplication*, which to a mathematics student is deceptively simple. (Perhaps a term like "elliptic multiplication" would be better, but here as elsewhere, the jargon has solidified.) This theory is informally described as an order of magnitude harder than the Kronecker–Weber theorem, and the case of *real* quadratic fields is more difficult still, and only partially understood.

SICs know a lot about Hilbert's twelfth problem [42]. They have been found to generate ray class fields over real quadratic extensions of the rationals $\mathbb{Q}(\sqrt{D})$, where D is the square-free part of the quantity $(d - 3)(d + 1)$. The path to counterparts for roots of unity goes through the *overlap phases,* the phases that the absolute-value bars in Eq. (1.1) throw away. The overlap phases turn out to be *algebraic integer units* in ray class fields or extensions thereof. (Roots of monic polynomials over $\mathbb{Z}$ are the algebraic integers, so called because their quotients yield the algebraic numbers just as quotients of ordinary integers $\mathbb{Z}$ yield the rationals $\mathbb{Q}$. The algebraic integers within a number field form a ring, and the units of this ring are those algebraic integers whose multiplicative inverses are also algebraic integers. In the real case, taking the absolute value discards a choice of ± 1, while in the complex case, it discards a phase that is not arbitrary, but rather a generalized kind of "± 1"!) Recently, Kopp has turned this connection around and, using conjectured properties of important numbers in ray class fields, constructed an exact SIC in $d = 23$ where none had been known before [43]. Kopp's SIC is constructed from overlap phases calculated as Galois conjugates of square roots of *Stark units*. These numbers figure largely in the Stark conjectures, which pertain to generating ray class fields explicitly. The conceptual waters here run deep, yet more remarkable still is the fact that these beguiling questions of number theory are, by way of almost schoolchildish geometry, intricated with quantum physics.

The SICs in dimensions $d = 2$ and $d = 3$, along with the Hoggar-type SICs in $d = 8$, stand out from the others in several ways, earning them the designation of *sporadic SICs*. These will be the primary focus of this book. While the other SICs connect with number theory in the way we have just outlined, the sporadic SICs instead seem to communicate with combinatorial design theory, finite simple groups and the exceptional Lie algebras. It has also been somewhat easier to do physics-oriented calculations with the sporadic SICs. They lie, as it were, under the lamp-post of the quantum information literature. My suspicion is that the physics questions to which the other SICs are the answers are mostly questions we have not yet learned how to ask.

References

1. P.W.H. Lemmens, J.J. Seidel, Equiangular lines. J. Algebra **24**(3), 494–512 (1973)
2. J.H. van Lint, J.J. Seidel, Equiangular point sets in elliptic geometry. Proc. Nederl Akad. Wetensch Ser. A **69**, 335–48 (1966)
3. G. Greaves, J. Syatriadi, P. Yatsyna, Equiangular lines in low dimensional Euclidean spaces (2020). arXiv:2002.08085
4. A. Barg, W.H. Yu, New bounds for equiangular lines. Contemp. Math. **625**, 111–21 (2014)
5. G. Greaves, J.H. Koolen, A. Munemasa, F. Szöllősi, Equiangular lines in Euclidean spaces. J. Comb. Theory, Ser. A **138**, 208–235 (2016). https://doi.org/10.1016/j.jcta.2015.09.008
6. L. Manivel, Configurations of lines and models of Lie algebras. J. Algebra **304**(1), 457–86 (2006). https://doi.org/10.1016/j.jalgebra.2006.04.029
7. J.J. Seidel, A. Blokhuis, H.A. Wilbrink, J.P. Boly, C.P.M. van Hoesel, Graphs and association schemes, algebra and geometry. Technical report EUT-Report Vol. 83-WSK-02, Technische Hogeschool Eindhoven (1983). https://pure.tue.nl/ws/portalfiles/portal/1994552/253169.pdf
8. H. Cohn, A. Kumar, S.D. Miller, D. Radchenko, M. Viazovska, Universal optimality of the E_8 and Leech lattices and interpolation formulas (2019)
9. D.E. Taylor, Regular 2-graphs. Proc. Lond. Math. Soc. **s3-35**(2), 257–274 (1977). https://doi.org/10.1112/plms/s3-35.2.257
10. N.I. Gillespie, Equiangular lines, incoherent sets and quasi-symmetric designs (2018). arXiv:1809.05739
11. P.J. Cameron, Strongly regular graphs, in *Topics in Algebraic Graph Theory*, eds. by L.W. Beineke, R.J. Wilson (Cambridge University Press, Cambridge, 2004), pp. 203–221
12. M.A. Sustik, J.A. Tropp, I.S. Dhillon, R.W. Heath Jr., On the existence of equiangular tight frames. Linear Algebra Appl. **426**, 619–35 (2007). https://doi.org/10.1016/j.laa.2007.05.043
13. E.J. King, X. Tang, New upper bounds for equiangular lines by pillar decomposition. SIAM J. Discret. Math. **33**(4), 2479–2508 (2019)
14. I. Balla, F. Dräxler, P. Keevash, B. Sudakov, Equiangular lines and spherical codes in Euclidean space. Invent. Math. **211**(1), 179–212 (2018)
15. J.C. Tremain, Concrete constructions of real equiangular line sets (2008). arXiv:0811.2779
16. H.S.M. Coxeter, *Regular Complex Polytopes*, 2nd edn. (Cambridge University Press, Cambridge, 1991)
17. G. Zauner, Quantendesigns. Grundzüge einer nichtkommutativen Designtheorie. Ph.D. thesis, University of Vienna (1999). https://doi.org/10.1142/S0219749911006776. http://www.gerhardzauner.at/qdmye.html; Published in English translation: G. Zauner, Quantum designs: foundations of a noncommutative design theory. Int. J. Quantum Inf. **9**, 445–508 (2011)
18. J.M. Renes, R. Blume-Kohout, A.J. Scott, C.M. Caves, Symmetric informationally complete quantum measurements. J. Math. Phys. **45**, 2171–2180 (2004). https://doi.org/10.1063/1.1737053
19. I. Bengtsson, SICs: some explanations. Found. Phys. **50**, 1794–1808 (2020). https://doi.org/10.1007/s10701-020-00341-9

20. P. Horodecki, L. Rudnicki, K. Życzkowski, Five open problems in quantum information (2020). arXiv:2002.03233
21. SIC Solutions. UMass Boston (2020). http://www.physics.umb.edu/Research/QBism/
22. I. Bengtsson, K. Życzkowski, Discrete structures in Hilbert space, in *Geometry of Quantum States: An Introduction to Quantum Entanglement*, 2nd edn. (Cambridge University Press, Cambridge, 2017)
23. S. Waldron, *An Introduction to Finite Tight Frames* (Springer, Berlin, 2018). https://doi.org/10.1007/978-0-8176-4815-2. https://www.math.auckland.ac.nz/~waldron/Preprints/Frame-book/frame-book.html
24. M.A. Nielsen, I. Chuang, *Quantum Computation and Quantum Information*, 10th anniversary edn. (Cambridge University Press, Cambridge, 2010)
25. N.D. Mermin, Hidden variables and the two theorems of John Bell. Rev. Mod. Phys. **65**(3), 803–15 (1993). https://doi.org/10.1103/RevModPhys.65.803
26. V. Veitch, S.A.H. Mousavian, D. Gottesman, J. Emerson, The resource theory of stabilizer computation. New J. Phys. **16**, 013009 (2014). https://doi.org/10.1088/1367-2630/16/1/013009
27. J.B. DeBrota, C.A. Fuchs, B.C. Stacey, Symmetric informationally complete measurements identify the irreducible difference between classical and quantum systems. Phys. Rev. Res. **2**, 013074 (2020). https://doi.org/10.1103/PhysRevResearch.2.013074
28. J. Conway, S. Kochen, The free will theorem. Found. Phys. **36**, 1441–73 (2006). https://doi.org/10.1007/s10701-006-9068-6
29. H. Weyl, *The Theory of Groups and Quantum Mechanics* (Dover, Illinois, 1931). Translated from the German *Gruppentheorie und Quantenmechanik* (S. Hirzel, 1928) by H. P. Robertson
30. G.V.R. Born, The wide-ranging family history of Max Born. Notes Rec. R. Soc. Lond. **56**(2), 219–62 (2002)
31. W.A. Fedak, J.J. Prentis, The 1925 Born and Jordan paper 'On quantum mechanics'. Am. J. Phys. **77**, 128 (2009). https://doi.org/10.1119/1.3009634
32. H. Zhu, Quantum state estimation and symmetric informationally complete POMs. Ph.D. thesis, National University of Singapore (2012). http://scholarbank.nus.edu.sg/bitstream/handle/10635/35247/ZhuHJthesis.pdf
33. S.G. Hoggar, Two quaternionic 4-polytopes, in *The Geometric Vein: The Coxeter Festschrift*, eds. by C. Davis, B. Grünbaum, F.A. Sherk (Springer, Berlin, 1981). https://doi.org/10.1007/978-1-4612-5648-9_14
34. S.G. Hoggar, 64 lines from a quaternionic polytope. Geom. Dedicata. **69**, 287–289 (1998). https://doi.org/10.1023/A:1005009727232
35. A. Szymusiak, W. Słomczyński, Informational power of the Hoggar symmetric informationally complete positive operator-valued measure. Phys. Rev. A **94**, 012122 (2015). https://doi.org/10.1103/PhysRevA.94.012122
36. H. Zhu, Super-symmetric informationally complete measurements. Ann. Phys. (NY) **362**, 311–326 (2015). https://doi.org/10.1016/j.aop.2015.08.005
37. R.A. Wilson, *The Finite Simple Groups* (Springer, Berlin, 2009)
38. J.H. Conway, D. Smith, *On Quaternions and Octonions: Their Geometry, Arithmetic, and Symmetry* (A K Peters, Natick, 2003)
39. M. Viazovska, The sphere packing problem in dimension 8. Ann. Math. **185**(3), 991–1015 (2017)
40. A.J. Scott, M. Grassl, Symmetric informationally complete positive-operator-valued measures: a new computer study. J. Math. Phys. **51**, 042203 (2010). https://doi.org/10.1063/1.3374022
41. M. Appleby, H. Yadsan-Appleby, G. Zauner, Galois automorphisms of a symmetric measurement. Quantum Inf. Comput. **13**, 672–720 (2013)
42. M. Appleby, S. Flammia, G. McConnell, J. Yard, SICs and algebraic number theory. Found. Phys. **47**, 1042–59 (2017). https://doi.org/10.1007/s10701-017-0090-7
43. G.S. Kopp, SIC-POVMs and the Stark conjectures. International Mathematics Research Notices p. rnz153 (2019)

Chapter 2
Optimal Quantum Measurements

We have our first serious encounter with the idea that SICs are extremal, and indeed optimal, quantum measurements. After presenting a thought-experiment that makes clear the meaning of the irreducible difference between classical and quantum, we learn how to construct SICs as group orbits, and we meet the sporadics up close.

2.1 Introduction

Somewhere in our first course on quantum mechanics, mixed in with the oversimplified history and the solution of the same differential equation in a tedious sequence of boundary conditions, we each encounter the double-slit experiment. A typical presentation might say that a source fires electrons at a wall with two gaps, and a detector on the far side of the wall clicks or does not click with some probability that depends upon its position. To get a sense that anything intriguing is happening, we compare multiple different, mutually exclusive variations on this arrangement, like blocking one or the other of the gaps, or inserting a detector that makes a "which way" determination upon the electron, registering that it passed through the left gap or the right. Quantum interference, as we learn to call it, manifests in the relation between probabilities that apply to different laboratory setups. For example, if $P(x)$ is the probability that an electron-detector at x will click, and $P(L, x)$ and $P(R, x)$ are the probabilities of an intermediate detection at the left or at the right gap followed by a final detector at x clicking, then interference is the fact that

$$P(x) \neq P(L, x) + P(R, x). \tag{2.1}$$

None of the terms on either side are themselves unusual. They are each real-valued, indeed bounded in the unit interval, and behave quite like classical probabilities. The peculiarity is that they do not *mesh together* in the intuitive way.

When I have taught the double-slit experiment, it has elicited less surprise than when I have lectured on other quantum surprises like the violation of Bell inequalities. This is not a controlled comparison, of course; for starters, the most recent lecture on the double slit was much earlier in the morning. Yet perhaps there is a lesson to be learned regardless: Can we push the double-slit experiment further and make it illuminate the quantum mysteries in more detail?

Our first step will be to abstract the idea by one level. Rather than speaking of electrons specifically, we will imagine an arbitrary physical system, which we can choose to subject to either one of two laboratory protocols. In the first protocol, we send the system directly into a measuring device and obtain an outcome. In the second protocol, we pass the system through an intermediate measuring device before feeding it into the same instrument as in the first protocol. We will denote the possible outcomes of the device that appears in both protocols by $\{D_j\}$, and the outcomes of the intermediate measurement in the two-stage protocol will be written $\{R_i\}$. The D here stands for "data", and the R for "reference". For this is our major advance over the textbook double slit: We are going to take advantage of the fact that quantum theory allows for a *reference measurement,* an experiment with the property that having probabilities for its outcomes completely characterizes the preparation of a system.

Figure 2.1 portrays the situation. The bottom path, drawn with the solid-line arrow, indicates the first protocol; and the top path, drawn with the dashed-line arrows, represents the second protocol with its intermediate R measurement. When a physicist Alice confronts this situation, she can write a probability distribution

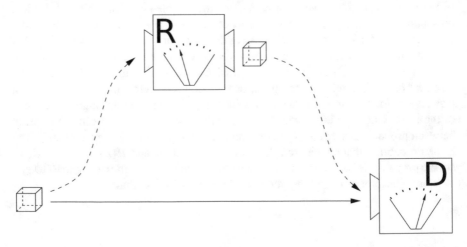

Fig. 2.1 Two alternative laboratory protocols, one with an intermediate measurement and the other without

$\{p(R_i) : i = 1, \ldots, N\}$ for the reference measurement's potential outcomes, and she can likewise write conditional probabilities $\{p(D_j|R_i)\}$ to capture her expectations about what might transpire in the second stage of the two-step protocol given a result of the reference measurement. The mathematical consistency requirements of probability theory then impose that

$$p(D_j) = \sum_{i=1}^{N} p(R_i)p(D_j|R_i) \,. \tag{2.2}$$

This relation is known, a trifle grandiosely, as the Law of Total Probability (LTP). But by itself, this Law cannot constrain Alice's probabilities for which D_j she might elicit by sending her system directly into that device *without* performing the reference measurement first. For clarity, we can distinguish these other probabilities by denoting them $\{q(D_j)\}$. If she forsakes the reference measurement, then she is meddling with the world in a different way, and the axioms of probability theory do not bind $p(D_j)$ with $q(D_j)$. *If* her theory of physics is such that she can honestly model the reference measurement as passively reading off an intrinsic property of the system, then she could write

$$q(D_j) = \sum_{i=1}^{N} p(R_i)p(D_j|R_i) \,, \tag{2.3}$$

but this is a statement of physics above and beyond mere mathematical self-consistency.

And indeed, it is *false* in quantum mechanics.

2.2 SIC Representations of Quantum States

A basic axiom of quantum theory is that to each physical system is associated a complex Hilbert space. In the domain of quantum information and computation, the Hilbert space is often taken to be finite-dimensional, with the dimension d scaling with the available budget. We mathematically represent a measurement by a positive operator valued measure, or POVM, which is a set of positive semidefinite operators on the system's Hilbert space that sum to the identity. Each operator in the set $\{E_i\}$ stands for an outcome of the measurement. When a physicist—let us call her Alice, per genre tradition—ascribes a quantum state to a system of interest, she writes a positive semidefinite operator of unit trace, i.e., a density matrix ρ. For a d-dimensional system (a *qudit*), ρ can be written as a $d \times d$ matrix of complex numbers. Alice's probability for obtaining the outcome E_i is given by the *Born rule*:

$$p(E_i) = \text{tr}(\rho E_i) \,. \tag{2.4}$$

The set of all valid density matrices ρ is a convex set whose extreme points are the rank-1 projection operators. These extreme points are also known as *pure states*; states that are not pure are designated *mixed*.

If the elements of a POVM span the space of Hermitian operators, then we can write any density matrix ρ as a linear combination of the POVM elements with real coefficients. This fact implies the possibility of *informationally complete* (IC) POVMs. Given a probability vector p over the outcomes of an IC POVM, we can reconstruct the density matrix ρ, and so we can in principle do anything we would have done with ρ using p instead. An IC POVM must have at least d^2 elements to span the operator space. A *minimal* IC POVM, or MIC, has exactly d^2 elements. MICs can be constructed in any dimension d [1, 2]; the question is how nice they can be made.

SICs are a special type of POVM. Given a set of d^2 equiangular unit vectors $\{|\pi_i\rangle\} \subset \mathbb{C}^d$, we can construct the operators which project onto them, and in turn we can rescale those projectors to form a set of POVM elements, or "effects":

$$E_i = \frac{1}{d}\Pi_i, \text{ where } \Pi_i = |\pi_i\rangle\langle\pi_i|. \tag{2.5}$$

The equiangularity condition on the $\{|\pi_i\rangle\}$ turns out to imply that the $\{\Pi_i\}$ are linearly independent, and thus they span the space of Hermitian operators on $\mathbb{C}^d$. Because the SIC projectors $\{\Pi_i\}$ form a basis for the space of Hermitian operators, we can express any quantum state ρ in terms of its (Hilbert–Schmidt) inner products with them. But, by the Born rule, the inner product $\text{tr}(\rho\Pi_i)$ is, apart from a factor $1/d$, just the probability of obtaining the ith outcome of the SIC measurement $\{E_i\}$. The formula for reconstructing ρ given these probabilities is quite simple, thanks to the symmetry of the projectors:

$$\rho = \sum_i \left[(d+1)p(i) - \frac{1}{d} \right] \Pi_i, \tag{2.6}$$

where $p(i) = \text{tr}(\rho E_i)$ by the Born rule. This furnishes us with a map from quantum state space into the probability simplex, a map that is one-to-one but not onto. In other words, we can fix a SIC as a reference measurement and then transform between density matrices and probability distributions without ambiguity, but the set of *valid* probability distributions for our reference measurement is a proper subset of the probability simplex.

We don't *need* equiangularity for informational completeness, just that the d^2 operators which form the reference measurement are linearly independent and thus span the operator space. But equiangularity implies the linear independence of those operators, and it makes the formula for reconstructing ρ from the overlaps particularly clean [3, 4].

SICs are sometimes discussed in the terminology of *frame theory*. A SIC provides a frame—more specifically, an *equiangular tight frame*—for the vector space $\mathbb{C}^d$. Given a finite-dimensional Hilbert space $\mathcal{H}$ with an inner product $\langle \cdot, \cdot \rangle$, a frame for $\mathcal{H}$

is a set of vectors $\{v_j\} \subset \mathcal{H}$ such that for any vector $u \in \mathcal{H}$,

$$A||u||^2 \le \sum_j |\langle v_j, u \rangle|^2 \le B||u||^2,\tag{2.7}$$

for some positive constants A and B. The frame is *equal-norm* if all the vectors $\{v_j\}$ have the same norm, and the frame is *tight* if the "frame bounds" A and B are equal. For more on this terminology and its history, we refer to Kovačević and Chebira [5, 6]. In our experience, the language of frames is more common among those who come to SICs from pure mathematics or from signal processing than among those motivated by quantum physics.

Calculating a density matrix ρ from the probabilities $p(i)$ is a special case of reconstructing a vector from its inner products with a set of frame vectors using a *dual frame* [7, 8]. The appeal of the SIC representation is that the reconstruction formula for ρ has such a simple form, in terms of physically meaningful quantities. One simply takes the probabilities, which are directly relevant to things the experimentalists can actually do [9–15], and shifts and rescales them to obtain the coefficients for the operators $\{\Pi_i\}$.

Because the projectors $\{\Pi_j\}$ enjoy such a nice symmetry property, we can *orthogonalize* them by shifting and rescaling, at the cost of losing positive semidefiniteness. In fact, there are two choices for doing so:

$$Q_j^\pm := \pm\sqrt{d+1}\,\Pi_j + \frac{1 \mp \sqrt{d+1}}{d} I.\tag{2.8}$$

The bases $\{Q_j^\pm\}$ have interesting properties with regard to Lie algebra theory [16] and the study of quantum probability [17–19].

Given any *other* POVM $\{D_j\}$, we can find its outcome probabilities by

$$q(D_j) := \mathrm{tr}(\rho D_j) = \sum_i \left[(d+1)p(R_i) - \frac{1}{d} \right] p(D_j|R_i),\tag{2.9}$$

where the conditional probability on the right-hand side is

$$p(D_j|R_i) := \mathrm{tr}(D_j \Pi_i).\tag{2.10}$$

Note that Eq. (2.9) has the form of the LTP

$$p(D_j) = \sum_i p(R_i)p(D_j|R_i),\tag{2.11}$$

but with the probabilities $p(R_i)$ "deformed" by a rescaling and a shift. In prior work, the importance of Eq. (2.9) was recognized by designating it the *urgleichung* ("primal equation" in German, or perhaps Klingon).

Historically, SICs were introduced as candidate reference measurements and the urgleichung was deduced. It stood out for being a simpler expression than, seemingly, had a right to exist. Later, this feeling was bolstered with theorems [20], the most general of which has been the following [4].

Let the reference measurement $\{R_i\}$ be an arbitrary MIC, and suppose that if a particular outcome R_i is registered, then the system is assigned a corresponding post-measurement state σ_i. In order to characterize the $\{D_j\}$ POVM, we need the states $\{\sigma_i\}$ to span the operator space like the $\{R_i\}$ do, but we do not need to restrict them otherwise. Each reference measurement has a *Born matrix* Φ, which is most conveniently defined through its inverse:

$$[\Phi^{-1}]_{ij} := \text{tr}(R_i \sigma_j) . \tag{2.12}$$

Because the post-measurements form a basis, we can write any density matrix ρ as

$$\rho = \sum_j \alpha_j \sigma_j , \tag{2.13}$$

where all the coefficients are real. The Born-rule probability of R_i is then

$$p(R_i) = \sum_j \alpha_j \text{tr}(R_i \sigma_j) = \sum_j [\Phi^{-1}]_{ij} \alpha_j . \tag{2.14}$$

Inverting this gives

$$\rho = \sum_i \left[\sum_k [\Phi]_{ik} p(R_k) \right] \sigma_i . \tag{2.15}$$

Consequently, the Born-rule probability of D_j is

$$q(D_j) = \sum_i \left[\sum_k [\Phi]_{ik} p(R_k) \right] p(D_j | R_i) . \tag{2.16}$$

The Born rule has the form of the Law of Total Probability, but with the matrix Φ inserted. We can write this more compactly by suppressing indices. If $p(R)$ without any subscript denotes the column vector whose entries are all the $\{p(R_i)\}$, and likewise for the other quantities in the previous equation, then

$$q(D) = p(D|R) \, \Phi \, p(R) . \tag{2.17}$$

This would coincide with the classical intuition to carry the LTP over from one protocol to the other, if it weren't for the Born matrix Φ there in the middle of it. The discrepancy between Φ and the identity matrix is an indicator of nonclassicality.

Let Φ_{SIC} denote the particular Born matrix appearing in the urgleichung (2.9). It can be proved [4] that Φ_{SIC} is the closest any Born matrix can be to the identity, with respect to any unitarily invariant norm. This is a large family of matrix norms, defined by the condition $||A|| = ||UAV||$ for all unitary matrices U and V, and it includes pretty much every notion of matrix norm that one has need of (Frobenius, Schatten, Ky Fan, et cetera). If $\{R_i\}$ is an arbitrary MIC and $\{\sigma_i\}$ an arbitrary basis of post-measurement states, then for any unitarily invariant norm $|| \cdot ||$,

$$||I - \Phi|| \geq ||I - \Phi_{\text{SIC}}|| \qquad (2.18)$$

with equality if and only if both $\{R_i\}$ and $\{\sigma_i\}$ are SICs. Consequently, SICs yield the *irreducible* nonclassicality of quantum theory: No reference measurement can bring the Born rule more in line with the Law of Total Probability.

The intuition at work here is that, classically, an informationally complete measurement would be, e.g., one that reads off a system's coordinates in phase space. Any other measurement would in principle be a coarse-graining of that information.[1] But in quantum theory, there is no underlying phase space, so we should not use a formula that depends upon the concept of one. By identifying this "minimum distance" between a probabilistic representation of quantum theory and classical probability, SICs provide a measure of exactly how nonclassical quantum physics is. This naturally raises the question of how this measure of nonclassicality relates to other such, of which the quintessential is the violation of a Bell inequality. We will return to this question in a later chapter.

The proof of the theorem that SICs provide the irreducible margin of nonclassicality is moderately involved, thanks to the fact that the matrices one must manipulate do not always enjoy the properties that quantum physicists are accustomed to matrices having. A special case of one direction of it is included in the Exercises. The measure-and-reprepare type of operation we have considered is also known as an *entanglement-breaking quantum channel* [24]. We refer to the literature for more on the relation between these and SICs [25, 26].

Note that the bracketed quantity in the urgleichung, $(d + 1)p(R_i) - 1/d$, can go negative if $p(R_i)$ is sufficiently small. This deformation of the vector of $p(R_i)$ is technically what is sometimes known as a "quasi-probability"—a vector whose sum is normalized to unity, but whose elements are not confined to the unit interval [7, 8, 17, 18, 27]. Negative "quasi-probabilities" are an artifact of trying to squeeze something into the form of the LTP that doesn't actually fit. Generally, states that are close to orthogonal to one of the SIC vectors will pick up negativity in what we might call their quasi-probability representation. But it's the $\{p(R_i)\}$ that are directly, operationally meaningful. There is an experiment that Alice could go into the lab and do, and $p(R_i)$ is how much she should bet on the ith outcome of it. Negativity of quasi-probability can become meaningful after one introduces a notion of "quasi-classical states", or "states that are easy to emulate on a classical machine". Once we

---

[1]Classical measurements can, of course, disturb the system that they are applied to. But this is a largely uninteresting complication. In order to express Pauli's "ideal of the detached observer" [21–23], reading off the system's intrinsic properties without disruption is clearly the correct idealization. Comparing the quantum and the classical in a reasonable way requires the plainest expression of both.

bring in the ideas necessary to support a "resource theory", then negativity can gain significance as a measure of how powerful a given resource is. But in the broader scheme of things, it is a secondary and somewhat incidental notion.[2]

Let us suppose we have a SIC solution for some dimension d. (In the next section, we will examine some examples in detail.) A state is pure if and only if its SIC representation satisfies the following two conditions. First, it must meet the quadratic constraint

$$\sum_j p(j)^2 = \frac{2}{d(d+1)}.$$ (2.19)

Second, it must satisfy the *QBic equation*,

$$\sum_{jkl} C_{jkl}\, p(j)\, p(k)\, p(l) = \frac{d+7}{(d+1)^3},$$ (2.20)

where we have introduced the *triple products,*

$$C_{jkl} = \operatorname{Re} \operatorname{tr}(\Pi_j \Pi_k \Pi_l).$$ (2.21)

If two or more indices are equal, this reduces to

$$\operatorname{tr}(\Pi_j \Pi_k) = \frac{d\delta_{jk}+1}{d+1}.$$ (2.22)

The set of all valid states is the convex hull of the probability distributions that satisfy Eqs. (2.19) and (2.20).

The quadratic constraint (2.19) has a considerably simpler structure than the QBic equation, so we investigate the former first. First, we take the reciprocal of both sides. Flipping this equation upside down makes both sides *look like counting!* The right-hand side becomes combinatorics: It's just the binomial coefficient for choosing two things out of $d+1$. Meanwhile, the left-hand side becomes the *effective number* of outcomes, which we are familiar with because it is a biodiversity index [28]:

$$N_{\text{eff}} = \left(\sum_i p(i)^2\right)^{-1}.$$ (2.23)

---

[2]Another reason to think of negativity as a secondary manifestation of nonclassicality is that we have considerable "gauge freedom" about where to put it. Adopting a vector notation for the urgleichung, we can express it as $q(D) = p(D|R)\, \Phi\, p(R)$, where Φ is a linear combination of the identity and the all-ones matrix [4]. In this form, it is clear that we can multiply Φ to the right, turning the vector $p(R)$ into "quasi-probabilities", or we could multiply Φ to the left, putting the negativity into the conditional probability matrix $p(D|R)$. We could even express Φ as $\Phi^{1/2}\Phi^{1/2}$ and split the negativity across both.

So, when we ascribe a pure state to a quantum system of Hilbert-space dimensionality d, we are saying that the effective number of possible outcomes for a reference measurement is

$$N_{\text{eff}} = \binom{d+1}{2}.$$ (2.24)

Consequently, ascribing a pure state means that we are effectively ruling out a number of outcomes equal to

$$d^2 - N_{\text{eff}} = \frac{d(d-1)}{2} = \binom{d}{2}.$$ (2.25)

This motivates the following question: What is the upper bound on the number of entries in **p** that can equal zero? A brief calculation with the Cauchy–Schwarz inequality [29] reveals that the answer is, in fact, exactly $d^2 - N_{\text{eff}}$. In other words, no vector **p** in the image of quantum state space can contain more than $d(d-1)/2$ zeros [30]. One reason why the sporadic SICs are distinguished from all the others is that they provide the examples where this bound is known to be saturated. We can attain it for three cases we will construct explicitly in the next section: qubit SICs (where it equals 1), the Hesse SIC (where it equals 3) and the Hoggar-type SICs (where it equals 28).

Since we have probability distributions, we can compute Shannon entropies. Of particular interest are the pure states which extremize the Shannon entropy of their SIC representations. It turns out (and the proof is not too long) that the pure states which *maximize* the Shannon entropy of their SIC representations are the SIC projectors $\{\Pi_i\}$ themselves.[3]

What about *minimizing* the Shannon entropy? Imagine a probability distribution, not necessarily one corresponding to a quantum state. Under the constraint that $\sum_i p(i)^2$ is fixed,

$$\sum_i p(i)^2 = \frac{1}{N},$$ (2.26)

then it can be shown [32] that the distributions of minimum entropy take the form

$$\left(\frac{1}{N}, \cdots, \frac{1}{N}, 0, \cdots, 0 \right).$$ (2.27)

Exactly N entries are nonzero, and the others all vanish. If we take

$$N = \frac{d(d+1)}{2} = \binom{d+1}{2},$$ (2.28)

then we see that a probability distribution with N nonvanishing, uniform entries is a pure state that minimizes the Shannon entropy—*provided* that it corresponds to a

---

[3]I first learned of this from unpublished notes by Huangjun Zhu, written in 2013. See also [31].

valid pure state. In other words, the minimizers we seek are those permutations of Eq. (2.27) that satisfy the QBic equation.

2.3 Constructing SICs Using Groups

When we count SICs in a dimension d, we do so up to unitary equivalence, since an overall unitary transformation of the entire set preserves the inner products between vectors. All known SICs have an additional kind of symmetry, above and beyond their definition: They are *group covariant*. Each SIC can be constructed by starting with a single vector, known as a *fiducial* vector, and acting upon it with the elements of some group. It is not known whether or not a SIC must be group covariant. Possibly, because group covariance simplifies the search procedure [33, 34], the fact that we only know of group-covariant SICs is merely an artifact. (However, we do have a proof that all SICs in $d = 3$ are group covariant [35].)

In all cases but one, namely the Hoggar SIC we will define below, the group that generates a SIC from a fiducial is an instance of a *Weyl–Heisenberg group*. We can define this group as follows. First, fix a value of d, and let $\omega_d = e^{2\pi i/d}$. Then, construct the shift and phase operators

$$X|j\rangle = |j + 1\rangle, \ Z|j\rangle = \omega_d^j|j\rangle,\tag{2.29}$$

where the shift is modulo d. These operators almost commute: Exchanging them introduces a factor of ω_d.

$$ZX = \omega_d XZ.\tag{2.30}$$

Consequently, together they generate a group whose elements have the form $\omega_d^\alpha X^m Z^n$, where all powers run from 0 to $d - 1$. For many purposes, those phase factors in front can be neglected, because they will cancel when we take the orbit of a fiducial projector Π_0 under conjugation.

In $d = 2$—that is, for a system comprising a single qubit—a SIC is simply a regular tetrahedron, inscribed in the Bloch sphere [36]. (This configuration was described by Feynman, in a 1987 festschrift for Bohm [27].) Let r and s be signs, and let σ_x, σ_y and σ_z denote the Pauli matrices. Then, the four pure states

$$\Pi_{r,s} = \frac{1}{2}\left(I + \frac{1}{\sqrt{3}}(r\sigma_x + s\sigma_y + rs\sigma_z)\right)\tag{2.31}$$

define a tetrahedron. Each point (x, y, z) lying within the unit ball (*Bloch ball*) defines a valid quantum state. The SIC representation of this state is the probability vector whose components are

$$p(r, s) = \frac{1}{4} + \frac{\sqrt{3}}{12}(sx + ry + srz).\tag{2.32}$$

Given the tetrahedron (2.31), we can define another, related to the first by inversion. Together, the two tetrahedra form a stellated octahedron, inscribed in the Bloch sphere. The SIC representations of the vertices of the second tetrahedron are the vector

$$\left(0, \frac{1}{3}, \frac{1}{3}, \frac{1}{3}\right) \tag{2.33}$$

and its permutations. Any tetrahedron inscribed in the Bloch sphere necessarily defines a qubit SIC; however, these two are "canonical" in that they play nicely with the choice of using the eigenvectors of Z to fix the basis in which we work.

Moving up a dimension, there is a SIC in $d = 3$ that is possibly even nicer than those in $d = 2$. It is known as the *Hesse SIC,* and we construct it by applying the Weyl–Heisenberg group to the fiducial

$$\left|\pi_0^{\text{(Hesse)}}\right\rangle := \frac{1}{\sqrt{2}}(0, 1, -1)^{\text{T}}. \tag{2.34}$$

This is just about as simple a vector as one could hope for, if one wanted to make something by applying the Weyl–Heisenberg group. There would be no fun in feeding in a basis vector like $(1, 0, 0)^{\text{T}}$, since X would just cycle it around and Z would only introduce physically irrelevant overall phases. So, we'd have no hope of building a full set of nine vectors. Instead, we turn to the next-flattest a vector can be!

Dimension $d = 3$, or *qutrit* state space, actually plays host to an uncountably infinite family of SICs, none of which can be rotated into each other. To the best of our knowledge, this only happens in $d = 3$, never in higher dimensions! To map this family, we introduce the concept of a *Clifford group.* Every Weyl–Heisenberg group has its Clifford group, which is the set of unitary operators that map the set of Weyl–Heisenberg operators to itself. In group-theory parlance, the Clifford group is the normalizer of the Weyl–Heisenberg group. It is convenient to think of the Weyl–Heisenberg operators as the points of a $d \times d$ grid, with powers of X along one axis and powers of Z along the other. Conjugating by a Clifford unitary sends this grid to itself:

$$X^m Z^n \rightarrow X^{m'} Z^{n'}, \tag{2.35}$$

with the map from (m, n) to (m', n') being a linear transformation of determinant 1.[4] It is not too difficult to show that applying a Clifford unitary to a Weyl–Heisenberg SIC fiducial always yields another Weyl–Heisenberg SIC fiducial—possibly another element in the same SIC, or possibly an element of a different SIC.

To construct the infinite family of qutrit SICs, we introduce a parameter $t \in [0, \frac{\pi}{6}]$ and the fiducials

$$|\pi_0(t)\rangle = \frac{1}{\sqrt{2}}(0, 1, -e^{2it})^{\text{T}}. \tag{2.36}$$

---

[4]Strictly speaking, the grid picture works when the dimension is *odd.* Extra complications occur with Clifford groups in even d, which are among those frustrating glitches in the cosmos for which nobody seems to have a good story yet. For our purposes, we will only need $d = 3$.

We then take the Clifford orbit of each vector. When $t = 0$, we recover the Hesse fiducial, and its Clifford orbit turns out to be just the Hesse SIC. At the other extreme, when $t = \frac{\pi}{6}$, the Clifford orbit of $\left|\pi_0(\frac{\pi}{6})\right\rangle$ is a set of *four* SICs. Due to their role as "magic states" in quantum computation, these 36 states are sometimes known as *Norrell states* [37].

Fiducials with different values of t give Clifford orbits that do not overlap and cannot be mapped to each other by Clifford unitaries. Zhu has proved that the SICs with paremeter values t, $\frac{\pi}{9} - t$ and $\frac{\pi}{9} + t$ can be mapped to each other using a non-Clifford unitary, but there are no other equivalences [38]. Consequently, any pair of SICs on orbits with $t, t' \in [0, \frac{\pi}{18}]$ are unitarily inequivalent.

These taxonomies of qubit and qutrit SICs are known to be complete. No other SICs, group covariant or otherwise, exist in these dimensions. The proof for $d = 3$ involves mathematics that is likely somewhat esoteric to a physics student, and moreover leans on computer assistance near the end [35]; I wonder if a simpler proof is possible.

The last set of sporadic SICs live in $d = 8$ and are designated the *Hoggar-type SICs*. The construction was first devised by Hoggar [39, 40] by starting with 64 nonequiangular diagonals through the vertices of a quaternionic polytope, which become 64 equiangular lines when converted to complex space. Hoggar's result was among the first discoveries of a maximal set of complex equiangular lines [41, pp. 731–33]. Actually, we have multiple choices of fiducial in this case, yielding distinct sets of $d^2 = 64$ states. However, all of these 240 sets have the same symmetry group, and they are equivalent to one another up to unitary or antiunitary transformations. For brevity, then, we will sometimes refer to "the" Hoggar SIC if ambiguity does not arise [42].

A fiducial for a Hoggar-type SIC [43] can be written as follows:

$$\left|\pi_0^{(\mathrm{Hoggar})}\right\rangle \propto (-1 + 2i, 1, 1, 1, 1, 1, 1, 1)^{\mathrm{T}}. \qquad (2.37)$$

Upon this, we act with the elements of the group that is the tensor product of three copies of the $d = 2$ Weyl–Heisenberg group:

$$k = (k_0, k_1, \ldots, k_5), \quad D_k = X^{k_0} Z^{k_1} \otimes X^{k_2} Z^{k_3} \otimes X^{k_4} Z^{k_5}. \qquad (2.38)$$

References

1. C.M. Caves, C.A. Fuchs, R. Schack, Unknown quantum states: the quantum de Finetti representation. J. Math. Phys. **43**(9), 4537–59 (2002). https://doi.org/10.1063/1.1494475
2. D.M. Appleby, Symmetric informationally complete measurements of arbitrary rank. Opt. Spect. **103**, 416–428 (2007). https://doi.org/10.1134/S0030400X07090111
3. M. Appleby, C.A. Fuchs, B.C. Stacey, H. Zhu, Introducing the Qplex: a novel arena for quantum theory. Eur. Phys. J. D **71**, 197 (2017). https://doi.org/10.1140/epjd/e2017-80024-y

4. J.B. DeBrota, C.A. Fuchs, B.C. Stacey, Symmetric informationally complete measurements identify the irreducible difference between classical and quantum systems. Phys. Rev. Res. **2**, 013074 (2020). https://doi.org/10.1103/PhysRevResearch.2.013074

5. J. Kovačević, A. Chebira, Life beyond bases: the advent of frames (part 1). IEEE Signal Process. Mag. **24**(4), 86–104 (2007). https://doi.org/10.1109/MSP.2007.4286567

6. J. Kovačević, A. Chebira, Life beyond bases: the advent of frames (part 2). IEEE Signal Process. Mag. **24**(5), 115–25 (2007). https://doi.org/10.1109/MSP.2007.904809

7. C. Ferrie, J. Emerson, Frame representations of quantum mechanics and the necessity of negativity in quasi-probability representations. J. Phys. A **41**, 352001 (2008). https://doi.org/10.1088/1751-8113/41/35/352001

8. C. Ferrie, J. Emerson, Framed Hilbert space: hanging the quasi-probability pictures of quantum theory. New J. Phys. **11**, 063040 (2009). https://doi.org/10.1088/1367-2630/11/6/063040

9. J. Du, M. Sun, X. Peng, T. Durt, Realization of entanglement assisted qubit-covariant symmetric informationally complete positive operator valued measurements. Phys. Rev. A **74**, 042341 (2006). https://doi.org/10.1103/PhysRevA.74.042341

10. T. Durt, C. Kurtsiefer, A. Lamas-Linares, A. Ling, Wigner tomography of two-qubit states and quantum cryptography. Phys. Rev. A **78**, 042338 (2008). https://doi.org/10.1103/PhysRevA.78.042338

11. Z.E.D. Medendorp, F.A. Torres-Ruiz, L.K. Shalm, G.N.M. Tabia, C.A. Fuchs, A.M. Steinberg, Experimental characterization of qutrits using symmetric informationally complete positive operator-valued measurements. Phys. Rev. A **83**(5), 051801(R) (2011). https://doi.org/10.1103/PhysRevA.83.051801

12. Z. Bian, J. Li, H. Qin, X. Zhan, R. Zhang, B.C. Sanders, P. Xue, Realization of single-qubit positive-operator-valued measurement via a one-dimensional photonic quantum walk. Phys. Rev. A **114**, 203602 (2015). arXiv:1501.05540

13. Y.Y. Zhao, N. Yu, P. Kurzyński, G.Y. Xiang, C.F. Li, G.C. Guo, Experimental realization of generalized qubit measurements based on quantum walks. Phys. Rev. A **91**(4), 042101 (2015). https://doi.org/10.1103/PhysRevA.91.042101

14. N. Bent, H. Qassim, A.A. Tahir, D. Sych, G. Leuchs, L.L. Sánchez-Soto, E. Karimi, R.W. Boyd, Experimental realization of quantum tomography of photonic qudits via symmetric informationally complete positive operator-valued measures. Phys. Rev. X **5**, 041006 (2015). https://doi.org/10.1103/PhysRevX.5.041006

15. Z. Li, H. Zhang, H. Zhu, Implementation of generalized measurements on a qudit via quantum walks. Phys. Rev. A **99**, 062342 (2019). https://doi.org/10.1103/PhysRevA.99.062342

16. D.M. Appleby, C.A. Fuchs, H. Zhu, Group theoretic, Lie algebraic and Jordan algebraic formulations of the SIC existence problem. Quantum Inf. Comput. **15**, 61–94 (2015)

17. H. Zhu, Quasiprobability representations of quantum mechanics with minimal negativity. Phys. Rev. Lett. **117**(12), 120404 (2016). https://doi.org/10.1103/PhysRevLett.117.120404

18. J.B. DeBrota, C.A. Fuchs, Negativity bounds for Weyl-Heisenberg quasiprobability representations. Found. Phys. **47**, 1009–30 (2017)

19. J.B. DeBrota, B.C. Stacey, Discrete Wigner functions from informationally complete quantum measurements. Phys. Rev. A **102**, 032221 (2020). https://doi.org/10.1103/PhysRevA.102.032221

20. J.B. DeBrota, C.A. Fuchs, B.C. Stacey, The varieties of minimal tomographically complete measurements. Int. J. Quantum Inf. (online before print) (2020). https://doi.org/10.1142/S0219749920400055

21. S. Gieser, *The Innermost Kernel: Depth Psychology and Quantum Physics*, Wolfgang Pauli's Dialogue with C. G. Jung (Springer, Berlin, 2005)

22. H. Atmanspacher, H. Primas (eds.), *Recasting Reality: Wolfgang Pauli's Philosphical Ideas and Contemporary Science* (Springer, Berlin, 2009)

23. C.A. Fuchs, Notwithstanding Bohr, the reasons for QBism. Mind Matter **15**(2), 245–300 (2017)

24. M.B. Ruskai, Qubit entanglement breaking channels. Rev. Math. Phys. **15**, 643–62 (2003). https://doi.org/10.1142/S0129055X03001710

25. S.K. Pandey, V.I. Paulsen, J. Prakash, M. Rahaman, Entanglement breaking rank and the existence of SIC POVMs. J. Math. Phys. **61**, 042203 (2020). https://doi.org/10.1063/1.5045184

26. J.B. DeBrota, B.C. Stacey, Lüders channels and the existence of Symmetric Informationally Complete measurements. Phys. Rev. A **100**, 062327 (2019). https://doi.org/10.1103/PhysRevA.100.062327

27. R.P. Feynman, Negative probability, in *Quantum Implications: Essays in Honour of David Bohm* (Routledge, Milton, 1987), pp. 235–248. http://cds.cern.ch/record/154856

28. T. Leinster, C.A. Cobbold, Measuring diversity: the importance of species similarity. Ecology **93**(3), 477–489 (2012). https://doi.org/10.1890/10-2402.1. http://johncarlosbaez.wordpress.com/2011/11/07/measuring-biodiversity/

29. B.C. Stacey, SIC-POVMs and compatibility among quantum states. Mathematics **4**(2), 36 (2016). https://doi.org/10.3390/math4020036

30. D.M. Appleby, Å. Ericsson, C. Fuchs, Properties of QBist state spaces. Found. Phys. **41**, 564–579 (2011). https://doi.org/10.1007/s10701-010-9458-7

31. A. Szymusiak, Pure states that are 'most quantum' with respect to a given POVM (2017). arXiv:1701.01139

32. A. Szymusiak, W. Słomczyński, Informational power of the Hoggar symmetric informationally complete positive operator-valued measure. Phys. Rev. A **94**, 012122 (2015). https://doi.org/10.1103/PhysRevA.94.012122

33. A.J. Scott, M. Grassl, Symmetric informationally complete positive-operator-valued measures: a new computer study. J. Math. Phys. **51**, 042203 (2010). https://doi.org/10.1063/1.3374022

34. M. Appleby, S. Flammia, G. McConnell, J. Yard, Generating ray class fields of real quadratic fields via complex equiangular lines (2016). arXiv:1604.06098

35. L.P. Hughston, S.M. Salamon, Surveying points in the complex projective plane. Adv. Math. **286**, 1017–1052 (2016). https://doi.org/10.1016/j.aim.2015.09.022

36. J.M. Renes, R. Blume-Kohout, A.J. Scott, C.M. Caves, Symmetric informationally complete quantum measurements. J. Math. Phys. **45**, 2171–2180 (2004). https://doi.org/10.1063/1.1737053

37. V. Veitch, S.A.H. Mousavian, D. Gottesman, J. Emerson, The resource theory of stabilizer computation. New J. Phys. **16**, 013009 (2014). https://doi.org/10.1088/1367-2630/16/1/013009

38. H. Zhu, SIC POVMs and Clifford groups in prime dimensions. J. Phys. A **43**, 305305 (2010)

39. S.G. Hoggar, Two quaternionic 4-polytopes, in *The Geometric Vein: The Coxeter Festschrift*, eds. by C. Davis, B. Grünbaum, F.A. Sherk (Springer, Berlin, 1981). https://doi.org/10.1007/978-1-4612-5648-9_14

40. S.G. Hoggar, 64 lines from a quaternionic polytope. Geom. Dedicata. **69**, 287–289 (1998). https://doi.org/10.1023/A:1005009727232

41. C.A. Fuchs, My struggles with the block universe (2014). arXiv:1405.2390

42. H. Zhu, Quantum state estimation and symmetric informationally complete POMs. Ph.D. thesis, National University of Singapore (2012). http://scholarbank.nus.edu.sg/bitstream/handle/10635/35247/ZhuHJthesis.pdf

43. J. Jedwab, A. Wiebe, A simple construction of complex equiangular lines, in *Algebraic Design Theory and Hadamard Matrices* (Springer, Berlin, 2015), pp. 159–169. https://doi.org/10.1007/978-3-319-17729-8_13

Chapter 3
Geometry and Information Theory for Qubits and Qutrits

3.1 Qubits

Because we can treat quantum states as probability distributions, we can apply the concepts and methods of probability theory to them, including Shannon's theory of information. The structures that I will discuss in the following sections came to my attention thanks to Shannon theory. In particular, the question of recurring interest is, "Out of all the extremal states of quantum state space—i.e., the 'pure' states $\rho = |\psi\rangle\langle\psi|$—which *minimize* the Shannon entropy of their probabilistic representation?" I will focus on the cases of dimensions 2, 3 and 8, where the so-called sporadic SICs occur. In these cases, the information-theoretic question of minimizing Shannon entropy leads to intricate geometrical structures.

Any time we have a vector in $\mathbb{R}^3$ of length 1 or less, we can map it to a 2×2 Hermitian matrix by the formula

$$\rho = \frac{1}{2}\left(I + x\sigma_x + y\sigma_y + z\sigma_z\right), \tag{3.1}$$

where (x, y, z) are the Cartesian components of the vector and $(\sigma_x, \sigma_y, \sigma_z)$ are the Pauli matrices. This yields a positive semidefinite matrix ρ with trace equal to 1; when the vector has length 1, we have $\rho^2 = \rho$, and the density matrix is a rank-1 projector that can be written as $\rho = |\psi\rangle\langle\psi|$ for some vector $|\psi\rangle$.

Given any polyhedron of unit radius or less in $\mathbb{R}^3$, we can feed its vertices into the Bloch representation and obtain a set of density operators (which are pure states if they lie on the surface of the Bloch sphere). For a simple example, we can do a regular tetrahedron. Let s and s' take the values ± 1, and define

$$\rho_{s,s'} = \frac{1}{2}\left(I + \frac{1}{\sqrt{3}}(s\sigma_x + s'\sigma_y + ss'\sigma_z)\right). \tag{3.2}$$

B. C. Stacey, *A First Course in the Sporadic SICs*,
SpringerBriefs in Mathematical Physics 41,
https://doi.org/10.1007/978-3-030-76104-2_3

To make these density matrices into a POVM, scale them down by the dimension. That is, take

$$E_{s,s'} = \frac{1}{2}\rho_{s,s'}. \tag{3.3}$$

Then, the four operators $E_{s,s'}$ will sum to the identity. In fact, they comprise a SIC.
   By introducing a sign change, we can define another SIC,

$$\tilde{\rho}_{s,s'} = \frac{1}{2}\left(I + \frac{1}{\sqrt{3}}(s\sigma_x + s'\sigma_y - ss'\sigma_z)\right). \tag{3.4}$$

Each state in the original SIC is orthogonal to exactly one state in the second. In the Bloch sphere representation, orthogonal states correspond to *antipodal* points, so taking the four points that are antipodal to the vertices of our original tetrahedron forms a second tetrahedron. Together, the states of the two SICs form a cube inscribed in the Bloch sphere.

   Here we have our first appearance of Shannon theory entering the story. With respect to the original SIC, the states $\{|\tilde{\pi}_i\rangle\}$ of the antipodal SIC all minimize the Shannon entropy. The two interlocking tetrahedra are, entropically speaking, dual structures.

   In summary: Given a tetrahedral SIC, we can define a SIC representation of state space. Minimizing the Shannon entropy over pure states, as we discussed earlier, yields the four states of the counterpart tetrahedron. Performing the same procedure with the Hesse SIC, we will find that the pure states that minimize the Shannon entropy are twelve in number. It is to that structure which we now turn.

3.2   Qutrits

A quantum system for which $d = 3$ is known as a *qutrit*. In $d = 3$, we can simplify the QBic Eq. (2.20) considerably, using the Hesse SIC [1–3]. In the Hesse SIC representation of qutrit state space, the QBic equation can be reduced to

$$\sum_j p(j)^3 - 3 \sum_{(j,k,l)\in S} p(j)p(k)p(l) = 0, \tag{3.5}$$

where the list S is a set of index triples (j, k, l) which can be constructed as the lines in a certain nine-point geometry [1–3]. We will explore this handy fact in depth momentarily; it is a consequence of the triple products of the Hesse SIC states taking a simple form. In turn, the structure of the triple products simplifies because the Hesse SIC has the property that its symmetry group acts *doubly transitively*. This is a kind of symmetry beyond the definition of a SIC and beyond group covariance: Using unitary operators that map the Hesse SIC to itself, we can send any pair of states in the Hesse SIC to any other.

Zhu has proved [4] that the only SICs whose symmetry groups act doubly transitively are the tetrahedral SICs in $d = 2$, the Hesse SIC in $d = 3$ and the Hoggar SIC in $d = 8$. In $d = 2$, the QBic equation simplifies so far that it becomes redundant, and the quadratic constraint is sufficient to define the state space. As we have seen, the QBic equation also simplifies for the Hesse SIC, in a way that brings discrete geometry into the picture. It is reasonable to guess that something similar will happen in dimension $d = 8$. Because the results are more intricate, we defer discussion until Chap. 5.

In dimension $d = 3$, we encounter a veritable cat's cradle of vectors [3]. First, there's the Hesse SIC. Like all informationally complete POVMs, it defines a probabilistic representation of quantum state space, in this case mapping from 3×3 density matrices to the probability simplex for 9-outcome experiments. As suggested earlier, we can look for the pure states whose probabilistic representations minimize the Shannon entropy. The result is a set of twelve states, which sort themselves into four orthonormal bases of three states apiece. What's more, these bases are *mutually unbiased*: The Hilbert–Schmidt inner product of a state from one basis with any state from another is always constant. In a sense, the Hesse SIC has a "dual" structure, and that dual is a set of Mutually Unbiased Bases (MUB). Two orthonormal bases are mutually unbiased if each vector in one has the same absolute inner product with each vector in the other. No more than $d + 1$ bases can all be mutually unbiased with respect to one another in dimension d. The construction from the Hesse SIC yields a complete set of 4 MUB. This duality relation is rather intricate: Each of the 9 SIC states is orthogonal to exactly 4 of the MUB states, and each of the MUB states is orthogonal to exactly 3 SIC states [3].

An easy way to remember these relationships is to consider the finite affine plane on nine points. This configuration is also known as the discrete affine plane on nine points, and as the Steiner triple system of order 3. That's a lot of different names for something which is pretty easy to put together! To construct it, first draw a 3×3 grid of points, and label them consecutively:

$$
\begin{array}{ccc}
1 & 2 & 3 \\
4 & 5 & 6 \\
7 & 8 & 9
\end{array}
\tag{3.6}
$$

These will be the points of our discrete geometry. To obtain the lines, we read along the horizontals, the verticals and the leftward and rightward diagonals:

$$
\begin{array}{ccc}
(123) & (456) & (789) \\
(147) & (258) & (369) \\
(159) & (267) & (348) \\
(168) & (249) & (357)
\end{array}
\tag{3.7}
$$

Each point lies on four lines, and every two lines intersect in exactly one point. For our purposes today, each of the points corresponds to a SIC vector, and each of the lines corresponds to a MUB vector, with point-line incidence implying orthogonality. The four bases are the four ways of carving up the plane into parallel lines (horizontals, verticals, diagonals and other diagonals).

Earlier, we defined the triple products C_{jkl} as the real part of the trace of the product of three SIC projectors. This definition implies that C_{jkl} is invariant under cyclic shifts of the indices, thanks to the cyclicity of the trace, and also invariant under swaps, which change the imaginary part but not the real. Moreover, conjugating all the projectors in a SIC by the same unitary does not change the triple products. The triple products of the Hesse SIC can be expressed geometrically: For three distinct projectors Π_j, Π_k and Π_l, the triple product C_{jkl} equals $-\frac{1}{8}$ when the points (j, k, l) lie on the same line, and $\frac{1}{16}$ whenever they do not. All those ways we can send the 3×3 grid to itself by Clifford operations have to leave the triple products invariant, so the possibilities are squeezed down quite dramatically. This is what lies behind the simplification of the QBic Eqs. (2.20) to (3.5).

To construct a MUB vector, pick one of the 12 lines we constructed above, and insert zeroes into those slots of a 9-entry probability distribution, filling in the rest uniformly. For example, picking the line $(1, 2, 3)$, we construct the probability distribution

$$\left(0, 0, 0, \frac{1}{6}, \frac{1}{6}, \frac{1}{6}, \frac{1}{6}, \frac{1}{6}, \frac{1}{6}\right). \tag{3.8}$$

This represents a pure quantum state that is orthogonal to the quantum state

$$\left(\frac{1}{6}, \frac{1}{6}, \frac{1}{6}, 0, 0, 0, \frac{1}{6}, \frac{1}{6}, \frac{1}{6}\right) \tag{3.9}$$

and to

$$\left(\frac{1}{6}, \frac{1}{6}, \frac{1}{6}, \frac{1}{6}, \frac{1}{6}, \frac{1}{6}, 0, 0, 0\right), \tag{3.10}$$

while all three of these have the same Hilbert–Schmidt inner product with the quantum state represented by

$$\left(0, \frac{1}{6}, \frac{1}{6}, 0, \frac{1}{6}, \frac{1}{6}, 0, \frac{1}{6}, \frac{1}{6}\right), \tag{3.11}$$

for example.

Let the projectors onto the 9 SIC vectors be Π_1 through Π_9. We can uniquely identify each of the projectors onto the MUB vectors by the three SIC vectors to which they are orthogonal. For example, M_{123} is orthogonal to Π_1, Π_2 and Π_3. The 12 MUB states are then

$$\begin{aligned} &M_{123}, \ M_{456}, \ M_{789}; \\ &M_{147}, \ M_{258}, \ M_{369}; \\ &M_{159}, \ M_{267}, \ M_{348}; \\ &M_{168}, \ M_{249}, \ M_{357}; \end{aligned} \tag{3.12}$$

where each row corresponds to an orthonormal basis of $\mathbb{C}^3$. These are the only states whose probabilistic representation has 3 zeros and is elsewhere flat which are allowed by the constraint (3.5).

3.3 Coherence

In practice it may be helpful to think of a given set of quantum states as representing laboratory procedures that are easy to do. If the vectors comprising one orthonormal basis represent preparations that are convenient or inexpensive, then it is reasonable to say that density matrices that are strongly off-diagonal in that basis correspond to preparations that are more costly. We might then ask, for example, what tasks become practical if we can carry out one costly preparation and an arbitrarily large number of cheap transformations [5]. Having put ourselves in the mindset of viewing off-diagonality as a *resource,* we encounter a natural generalization: What if we have a choice of inexpensive bases? For example, we might be dealing with transmission errors that stochastically flip which basis is cheap [6]. Are there preparation procedures that are *equally costly* with respect to any one of a canonical discrete set of bases?

To make the question concrete, take the case of a single qubit. A quantum state that can be ascribed to a qubit-sized system is a 2×2 positive semidefinite matrix of unit trace, which we can neatly express as a sum over Pauli operators:

$$\rho = \frac{1}{2}(I + x\sigma_x + y\sigma_y + z\sigma_z) = \frac{1}{2}\begin{pmatrix} 1+z & x-iy \\ x+iy & 1-z \end{pmatrix}. \tag{3.13}$$

Here, (x, y, z) are the coordinates of the state ρ in the Bloch sphere representation. A handy measure of how off-diagonal the state ρ is in the eigenbasis of the Pauli operator σ_z is the sum of the squared magnitudes of the off-diagonal entries, which is

$$\frac{1}{4}|x - iy|^2 + \frac{1}{4}|x + iy|^2 = \frac{1}{2}(x^2 + y^2). \tag{3.14}$$

What states are equally off-diagonal by this measure in the eigenbases of σ_x, σ_y and σ_z? The coordinates of such a state must satisfy

$$x^2 + y^2 = x^2 + z^2 = y^2 + z^2. \tag{3.15}$$

Let us confine our attention to pure states, which lie on the surface of the Bloch ball:

$$x^2 + y^2 + z^2 = 1. \tag{3.16}$$

Combining these constraints, we find that

$$x^2 = y^2 = z^2 = \frac{1}{3},$$ (3.17)

meaning that the states we seek are the vertices of a cube inscribed in the Bloch sphere:

$$x, y, z \in \left\{ \pm \frac{1}{\sqrt{3}} \right\}.$$ (3.18)

This set of eight states naturally breaks down into two sets of four, which are orbits under the action of the Pauli group. Each set of four states forms a tetrahedron inscribed in the Bloch sphere, with the vertices of one tetrahedron antipodal to those of the other. One tetrahedron comprises the sign choices of even parity, and the other the sign choices of odd parity. We conclude that the set of qubit pure states that are equicoherent in all three Pauli bases are the qubit SIC states.

Note that if we had used the "l_1-norm of coherence" [7] instead, we would have found the constraint

$$\sqrt{x^2 + y^2} = \sqrt{x^2 + z^2} = \sqrt{y^2 + z^2},$$ (3.19)

to ultimately the same effect.

In any finite dimension, pure quantum states satisfy

$$\text{tr}\rho = \text{tr}\rho^2 = \text{tr}\rho^3 = 1.$$ (3.20)

Since the trace of $\rho^2 = \rho^\dagger \rho$ is the sum of the squared magnitudes of all the elements of ρ, we can relate the above measure of off-diagonality to the diagonal entries:

$$\sum_{i \neq j} |\rho_{ij}|^2 = 1 - \sum_i |\langle i|\rho|i\rangle|^2.$$ (3.21)

The numbers $\{\langle i|\rho|i\rangle\}$ are, of course, the probabilities ascribed to the outcomes of a measurement in the orthonormal basis $\{|i\rangle\}$.

Suppose that the dimension d is a power of a prime, so that a complete set of $d + 1$ Mutually Unbiased Bases (MUB) is known to exist. Let $|m, j\rangle$ be the states comprising these MUB, with m labeling the basis and j the vector within that basis. Given a state ρ, we apply the Born rule to compute the probabilities

$$p_{m,j} = \langle m, j|\rho|m, j\rangle.$$ (3.22)

It is a property of MUB that if ρ is a pure state, then

$$\sum_{m,j} p_{m,j}^2 = 2.$$ (3.23)

A *minimum uncertainty state* [8] distributes this sum equally over all the $d + 1$ bases:

$$\sum_j p_{m,j}^2 = \frac{2}{d+1} \ \forall \ m. \tag{3.24}$$

We see that any minimum uncertainty state will be equally off-diagonal in all $d + 1$ of the MUB. This generalizes the result we found above, since the states of SICs that are generated as orbits of the Weyl–Heisenberg group are minimum uncertainty states [8].

The term "coherence" is rather drastically overloaded, having different meanings in multiple fields, with SICs being important for many of them. They are significant for "coherence" in the Dutch-book and frame-theoretic senses of the word [9–11], and now we see that they are so in the "quantum coherence as a resource" sense as well.

The literature is replete with alternative ways of quantifying the off-diagonality of quantum states. (My impression is that some definitions lead to measures that might have more physical relevance, while others are easier to calculate, and sometimes the practical thing to do is try and use the latter to get a bound on the former.) Another such measure has an information-theoretic flavor and is known as the *relative entropy of coherence*. First, we define a "dephasing" operator that Procrusteanizes a density operator into a basis:

$$\Delta(\rho) = \sum_i (\langle i | \rho | i \rangle) |i\rangle\langle i|. \tag{3.25}$$

The relative entropy of coherence for a state ρ is the change in von Neumann entropy between its original and dephased forms:

$$C_r(\rho) = S(\Delta(\rho)) - S(\rho). \tag{3.26}$$

We focus our attention on pure states, for which the latter term vanishes and the relative entropy of coherence reduces to a simple Shannon functional of the probabilities $\{\langle i | \rho | i \rangle\}$.

A *MUB-balanced* state is one for which the Born-rule probabilities for the measurements corresponding to different bases are the same up to permutations [12, 13]. For any bases m and m',

$$p_{m,j} = p_{m',j'} \tag{3.27}$$

for some index j'. The Shannon functional is indifferent to permutations of probability vectors, and so MUB-balanced states are equicoherent across the MUB with respect to the relative entropy of coherence. Wootters and Sussman demonstrated that MUB-balanced states are minimum uncertainty states [12], implying that they are also equicoherent with respect to the sum-of-squared-magnitudes definition of coherence. We can see this from Eq. (3.21), since the sum over squared probabilities does not depend upon their ordering.

In particular, the nine states of the Hesse SIC in $d = 3$ are all MUB-balanced. For each state in the Hesse SIC and each basis m, $p_{m,j}$ is some permutation of the tuple $(0, \frac{1}{2}, \frac{1}{2})$. The combinatorics and finite geometry that make this pattern possible also yield a Kochen–Specker proof for qutrits [14] and are relevant for identifying the Hesse SIC states as *maximally magic resources* for quantum computation [15].

The Hoggar-type SICs in dimension 8 are sets of 64 states constructed as orbits of the three-qubit Pauli group [16, 17]. A convenient starting point is the vector

$$|\pi_0\rangle \propto (-1 + 2i, 1, 1, 1, 1, 1, 1, 1)^{\mathrm{T}}. \tag{3.28}$$

Taking the orbit of this state under the three-qubit Pauli group yields a set of 64 equiangular complex lines. As we mentioned earlier, other choices of initial vector are possible, but all the SICs found as orbits of the three-qubit Pauli group are equivalent up to unitary or antiunitary conjugations. The SIC generated from $|\pi_0\rangle$ is the prototype on which we will focus our attention.

Written in terms of rank-1 projectors, Hoggar SIC states satisfy

$$\mathrm{tr}\Pi_j\Pi_k = \frac{1}{9} \tag{3.29}$$

whenever $j \neq k$. Like the qubit and Hesse SIC states, these have been identified as resources for quantum computation [18]. Without explicit calculation, we can already see that these states will display degeneracies among their coherences. The relative entropy of coherence for the state Π_0 is the von Neumann entropy of the "dephased" state

$$\Delta(\Pi_0) = \sum_{i=0}^{8} (\langle i|\Pi_0|i\rangle)|i\rangle\langle i|. \tag{3.30}$$

But if we use a canonical set of MUB, each of the vectors $\{|i\rangle\}$ is defined as a simultaneous eigenstate of multiple three-qubit Pauli operators [19, 20]. Therefore, if U is a three-qubit Pauli unitary of which $\{|i\rangle\}$ are eigenvectors,

$$\Delta(\Pi_0) = \sum_{i=0}^{8} (\langle i|U^\dagger\Pi_0 U|i\rangle)|i\rangle\langle i| = \Delta(U^\dagger\Pi_0 U). \tag{3.31}$$

Because U is an element in the same group whose action generates the SIC,

$$\Delta(\Pi_0) = \Delta(\Pi_j) \tag{3.32}$$

for some value of j. So, for each choice from the $d + 1 = 9$ MUB, seven other SIC states will "dephase" to the same mixed state as Π_0 does.

The group covariance of the SIC set implies that for any Π_j,

$$\Delta(\Pi_j) = \Delta(D_j\Pi_0 D_j^\dagger) \tag{3.33}$$

for some three-qubit Pauli operator D_j. The set of all unitaries that map a Hoggar-type SIC to itself is a subset of the three-qubit Clifford group. Moreover, the three-qubit Clifford group maps the MUB states to each other. The von Neumann entropy of the "dephased" state $S(\Delta(\rho))$ depends only upon the values $\{\langle i|\rho|i\rangle\}$. If we fix $|i'\rangle = D_j^\dagger|i\rangle$, then

$$\langle i|\Pi_0|i\rangle = \langle i'|\Pi_j|i'\rangle. \tag{3.34}$$

The relative entropy of coherence for Π_0 with respect to the basis $\{|i\rangle\}$ is thus equal to that for Π_j with respect to the basis $\{|i'\rangle\}$.

We might plausibly guess that any state in a Hoggar-type SIC will turn out to be equicoherent across all nine MUB as well, thanks to the large size of its stabilizer group [4, 21, 22]. That is, there are 6,048 different Clifford unitaries which map the set $\{\Pi_j\}$ to itself and satisfy $U\Pi_0 U^\dagger = \Pi_0$. Moreover, the symmetry group of a Hoggar-type SIC is *doubly transitive,* able to map any pair of elements into any other. Let U be a Clifford unitary in the stabilizer of Π_0, so that $U\Pi_0 U^\dagger = \Pi_0$ and $U\Pi_j U^\dagger = \Pi_k$. Because $\Pi_j = D_j\Pi_0 D_j^\dagger$ for some three-qubit Pauli operator D_j, then

$$U D_j \Pi_0 D_j^\dagger U^\dagger = D_k \Pi_0 D_k^\dagger. \tag{3.35}$$

But we can conjugate our state Π_0 by the stabilizer unitary U and regroup:

$$(U D_j U^\dagger)\Pi_0(U D_j^\dagger U^\dagger) = D_k \Pi_0 D_k^\dagger. \tag{3.36}$$

The only way it seems that this can work out is if D_k, which is both unitary and Hermitian, is the same operator as that gotten by conjugating D_j with U. From the double transitivity of the Hoggar symmetry group, it follows that there must be a unitary in the stabilizer of Π_0 that can turn any Π_j into any desired Π_k. This in turn appears to require that the stabilizer of Π_0 is transitive on the Pauli operators $\{D_j\}$.

The states of the Hoggar SIC are not MUB-balanced, but they are minimum-uncertainty. By directly checking the overlaps with the Wootters–Fields MUB states [23], we find that the probability distribution $\mathbf{p}_m$ is, for each basis, a permutation either of the vector

$$\left(\frac{5}{12}, \frac{1}{12}, \frac{1}{12}, \frac{1}{12}, \frac{1}{12}, \frac{1}{12}, \frac{1}{12}, \frac{1}{12}\right) \tag{3.37}$$

or of the vector

$$\left(\frac{1}{3}, \frac{1}{6}, \frac{1}{6}, \frac{1}{6}, \frac{1}{6}, 0, 0, 0\right). \tag{3.38}$$

These two vectors have the same 2-norm, and so

$$\sum_j p_{m,j}^2 = \frac{2}{9} \tag{3.39}$$

for all 9 choices of m. In fact, for each of the 64 Hoggar states, 2 of the bases yield the first vector, and the other 7 bases yield the second. This establishes equicoherence with respect to the definition (3.21). It also establishes equicoherence with respect to an information-theoretic measure like Eq. (3.26), if the entropy functional is the Rényi 2-entropy rather than the Shannon formula [24].

This example presents an intriguing generalization of the MUB-balanced state concept: "almost MUB-balanced" quantum states, where there are (up to permutations) two distinct probability vectors, both representing "equal uncertainty".

It also follows from the group covariance of a Hoggar-type SIC that

$$\langle \pi_j | D_k | \pi_j \rangle = \pm \frac{1}{3}, \tag{3.40}$$

where D_k is any three-qubit Pauli operator. Using these operators as a Hermitian basis, we can write any Π_j as a linear combination of them, and the magnitudes of the coefficients in the expansion will be the same for all j. A generalization of the equicoherence property from which we derived the qubit SIC states follows naturally.

To recap: Coherence, treated as a resource in quantum information theory, is a basis-dependent quantity. Looking for states that have constant coherence under canonical changes of basis yields highly symmetric structures in state space. For the case of a qubit, we find an easy construction of qubit SICs. Moreover, we find that SICs in dimension 3 and 8 are also equicoherent.

From one perspective, coherent superpositions are not the deepest of the quantum mechanical mysteries. It is possible to construct them in theories that have underlying local hidden variables and that offer no hope of computational speed-up. The idea in old books that interference effects are quintessentially nonclassical is, in a modern analysis, a failure of imagination [25–27]. Useful as coherence may be for some protocols, it does not appear to be the most potent resource within the scope of quantum theory. Equicoherence, on the other hand, takes us out of that intermediate, semiclassical regime. We will explore how this works in the next chapter.

References

1. G.N.M. Tabia, Experimental scheme for qubit and qutrit symmetric informationally complete positive operator-valued measurements using multiport devices. Phys. Rev. A **86**, 062107 (2012). https://doi.org/10.1103/PhysRevA.86.062107
2. G.N.M. Tabia, D.M. Appleby, Exploring the geometry of qutrit state space using symmetric informationally complete probabilities. Phys. Rev. A **88**(1), 012131 (2013). https://doi.org/10.1103/PhysRevA.88.012131
3. B.C. Stacey, SIC-POVMs and compatibility among quantum states. Mathematics **4**(2), 36 (2016). https://doi.org/10.3390/math4020036
4. H. Zhu, Super-symmetric informationally complete measurements. Ann. Phys. (NY) **362**, 311–326 (2015). https://doi.org/10.1016/j.aop.2015.08.005
5. A. Streltsov, G. Adesso, M.B. Plenio, Colloquium: quantum coherence as a resource. Rev. Mod. Phys. **89**, 041003 (2017). https://doi.org/10.1103/RevModPhys.89.041003

6. C.H. Bennett, C.A. Fuchs, J.A. Smolin, Entanglement-enhanced classical communication on a noisy quantum channel, in *Quantum Communication, Computing, and Measurement* (Springer, Berlin, 1997)

7. H. Zhu, M. Hayashi, L. Chen, Axiomatic and operational connections between the l_1-norm of coherence and negativity. Phys. Rev. A **97**, 022342 (2018). https://doi.org/10.1103/PhysRevA.97.022342

8. M. Appleby, H.B. Dang, C.A. Fuchs, Symmetric informationally-complete quantum states as analogues to orthonormal bases and minimum uncertainty states. Entropy **16**, 1484 (2014)

9. C.A. Fuchs, R. Schack, Quantum-Bayesian coherence. Rev. Mod. Phys. **85**, 1693–1715 (2013). https://doi.org/10.1103/RevModPhys.85.1693

10. B. Bodmann, J. Haas, A short history of frames and quantum designs (2017). arXiv:1709.01958

11. C.A. Fuchs, B.C. Stacey, QBism: quantum theory as a hero's handbook (2016). arXiv:1612.07308

12. W.K. Wootters, D.M. Sussman, Discrete phase space and minimum-uncertainty states (2007). arXiv:0704.1277

13. M. Appleby, I. Bengtsson, H.B. Dang, Galois unitaries, Mutually Unbiased Bases, and MUB-balanced states (2014). arXiv:1409.7887

14. I. Bengtsson, K. Blanchfield, A. Cabello, A Kochen-Specker inequality from a SIC. Phys. Lett. A **376**, 374–376 (2012). https://doi.org/10.1016/j.physleta.2011.12.011

15. V. Veitch, S.A.H. Mousavian, D. Gottesman, J. Emerson, The resource theory of stabilizer computation. New J. Phys. **16**, 013009 (2014). https://doi.org/10.1088/1367-2630/16/1/013009

16. S.G. Hoggar, 64 lines from a quaternionic polytope. Geom. Dedicata. **69**, 287–289 (1998). https://doi.org/10.1023/A:1005009727232

17. A. Szymusiak, W. Słomczyński, Informational power of the Hoggar symmetric informationally complete positive operator-valued measure. Phys. Rev. A **94**, 012122 (2015). https://doi.org/10.1103/PhysRevA.94.012122

18. E. Campbell, M. Howard, Application of a resource theory for magic states to fault-tolerant quantum computing. Phys. Rev. Lett. **118**, 090501 (2017). https://doi.org/10.1103/PhysRevLett.118.090501

19. J. Lawrence, C. Brukner, A. Zeilinger, Mutually unbiased binary observable sets on n qubits. Phys. Rev. A **65**, 032320 (2002). https://doi.org/10.1103/PhysRevA.65.032320

20. J.L. Romero, G. Björk, A.B. Klimov, L.L. Sánchez-Soto, On the structure of the sets of mutually unbiased bases for n qubits. Phys. Rev. A **72**, 062310 (2005). https://doi.org/10.1103/PhysRevA.72.062310

21. H. Zhu, Quantum state estimation and symmetric informationally complete POMs. Ph.D. thesis, National University of Singapore (2012). http://scholarbank.nus.edu.sg/bitstream/handle/10635/35247/ZhuHJthesis.pdf

22. B.C. Stacey, Sporadic SICs and the normed division algebras. Found. Phys. **47**, 1060–64 (2017). https://doi.org/10.1007/s10701-017-0087-2

23. W.K. Wootters, B.D. Fields, Optimal state-determination by mutually unbiased measurements. Ann. Phys. **191**, 363–81 (1989). https://doi.org/10.1016/0003-4916(89)90322-9

24. H. Zhu, M. Hayashi, L. Chen, Coherence and entanglement measures based on Rényi relative entropies. J. Phys. A **50**, 475303 (2017). https://doi.org/10.1088/1751-8121/aa8ffc

25. R.W. Spekkens, Evidence for the epistemic view of quantum states: a toy theory. Phys. Rev. A **75**(3), 032110 (2007). https://doi.org/10.1103/PhysRevA.75.032110

26. R.W. Spekkens, Quasi-quantization: classical statistical theories with an epistemic restriction, in *Quantum Theory: Informational Foundations and Foils*, eds. by G. Chiribella, R.W. Spekkens (eds.) (Springer, Berlin, 2016), pp. 83–135. https://doi.org/10.1007/978-94-017-7303-4_4

27. R.W. Spekkens, Reassessing claims of nonclassicality for quantum interference phenomena (2016). http://pirsa.org/16060102/

Chapter 4
SICs and Bell Inequalities

Close study of three sporadic SICs reveals an illuminating relation between different ways of quantifying the extent to which quantum theory deviates from classical expectations. In brief, the SIC outcome does not exist when no one is looking.

What feature of quantum physics distinguishes it from classical mechanics? Back in 1935, Schrödinger's answer was "entanglement" [1]. Today, though, this response is rather passé. On the one hand, we have learned that entanglement is not unique to quantum mechanics, but occurs rather generically in nonclassical theories that lack superluminal signalling [2, 3]. On the other hand, we know that the mere occurrence of entanglement in a theory is, quantifiably, less exotic than the violation of a Bell inequality [4–7]. And if our answer is "quantum phenomena can violate a Bell inequality", then a new question naturally arises. Can we *deduce* the structure of quantum theory *from* the violation of a Bell inequality? If we can't, then can we really say that we have put our finger on what the *central* mystery of quantum physics is?

This is a big question, and we can only claim to make partial progress towards answering it. But even the incomplete answers are intriguing, and at times, even evocative.

We will study this using examples in Hilbert spaces of decreasing dimension: first eight (three qubits), then four (two qubits) and finally three (a single qutrit). We begin with Mermin's three-qubit Bell inequality [8–11]. From there, we will turn to the Hoggar SIC [12–16], an eight-dimensional structure that provides a common meeting ground for two ways of discussing the nonclassicality of quantum theory. On the one hand, it furnishes a SIC representation of eight-dimensional quantum state space [17], and so it exemplifies the nonclassical meshing of probability assignments described by Fuchs and Schack [18–23]. On the other hand, that same state space is what one requires for the GHZ *gedankenexperiment* and for Mermin's three-qubit Bell inequality. Considering the Hoggar SIC will be enough to answer the title question

© The Author(s), under exclusive license to Springer Nature Switzerland AG 2021
B. C. Stacey, *A First Course in the Sporadic SICs*,
SpringerBriefs in Mathematical Physics 41,
https://doi.org/10.1007/978-3-030-76104-2_4

in the negative; we will then explore additional nuances by developing the theme using qubit and qutrit SICs. Finally, we will conclude with some thoughts on the project of reconstructing quantum theory from physical principles.

4.1   Mermin's Three-Qubit Bell Inequality

Let X, Y and Z denote the Pauli operators, and write XXX for $X \otimes X \otimes X$ and so forth. Then we can write Mermin's three-qubit Bell inequality [9, 10] in terms of a linear combination of expectation values:

$$B(\rho) = \langle XXX \rangle - \langle XYY \rangle - \langle YXY \rangle - \langle YYX \rangle. \tag{4.1}$$

One employs this inequality in the following manner. First, one argues that the hypothesis of local hidden variables implies

$$-2 \leq B(\rho) \leq 2. \tag{4.2}$$

A way to see why these bounds should be set at ± 2 is as follows. Suppose that each part of the tripartite system carries a pair of physical properties that respectively determine the outcomes of an X measurement and of a Y measurement performed on that part. As a whole, then, the system carries a set of properties

$$\lambda = (\lambda_{1X}, \lambda_{1Y}, \lambda_{2X}, \lambda_{2Y}, \lambda_{3X}, \lambda_{3Y}), \tag{4.3}$$

such that if we knew these values, we could say

$$\begin{aligned} \langle XXX \rangle - \langle XYY \rangle - \langle YXY \rangle - \langle YYX \rangle \\ = \lambda_{X1}\lambda_{X2}\lambda_{X3} - \lambda_{X1}\lambda_{Y2}\lambda_{Y3} - \lambda_{Y1}\lambda_{X2}\lambda_{Y3} - \lambda_{Y1}\lambda_{Y2}\lambda_{X3}. \end{aligned} \tag{4.4}$$

It is now a matter of arithmetic to verify that for each assignment of $+1$ and -1 to the six λs, this quantity is either $+2$ or -2. The list of values denoted by λ is, in older jargon, a "dispersion-free state" [24–27]. Since the sum of expectation values given any dispersion-free state is ± 2, any probabilistic average over dispersion-free states will lie in the interval $[-2, 2]$.

Having established the bounds in (4.2), one then finds a state—for example, the GHZ state—for which those bounds are violated. This establishes that quantum probabilities cannot be accounted for by local hidden variables. The GHZ state ρ_{GHZ} is (by definition [8]) an eigenstate of the operator XXX with eigenvalue $+1$, and it is also an eigenstate of XYY, of YXY and of YYX with eigenvalue -1. Therefore, $B(\rho_{\text{GHZ}}) = 4$. This means that ρ_{GHZ} violates the inequality (4.2), and thus, the statistics encapsulated in ρ_{GHZ} defy local classical emulation.

In discussions of hidden-variable models, it usually does not particularly matter what other mathematical structure the set of all λ's might have. The λ's might, for all we end up caring, be labeled by the elements of a group, or the morphisms of a groupoid, or the open sets of a topology; they could have any geometry, or none. (Indeed, it is fair to say that the nature of λ-space is "rarely subject to much critical scrutiny" [28].) What does matter is the hypothesis that each part of a system carries its part of λ with it as an intrinsic physical property. In the example above, we hypothesized that each of the three qubits carried its own, intrinsic λ_X and λ_Y. A preparation of the system naturally corresponds, then, to a probability distribution over the set of all possible λ's, or in other words, to a point in the simplex whose vertices are labeled by the possible values of λ. We could choose to decorate these vertices with additional structure (say, making them into a group), but that extra mathematical ornamentation is secondary to the *physical* assumption which makes our state space into a simplex and, ultimately, powers the derivation of Bell inequalities [25].

4.2 The Hoggar SIC

Consider the tensor product of three copies of qubit state space. We will take for our computational basis the tensor-product basis of Pauli Z eigenstates.

Now, we construct the Hoggar SIC, which will provide a "Bureau of Standards" experiment—a reference measurement with respect to which we can represent quantum theory in wholly probabilistic terms [29]. This construction is an example of how all known SICs are generated: We begin with a *fiducial vector* and take its orbit under the action of a group [30]. A convenient fiducial for our present purpose is the vector given up to normalization by

$$\left| \pi_0^{(\text{Hoggar})} \right\rangle \propto (-1 + 2i, 1, 1, 1, 1, 1, 1, 1)^{\text{T}}. \tag{4.5}$$

We apply the three-qubit Pauli group to generate the Hoggar SIC [14, 15, 31]. This is a set of 64 equiangular unit vectors $\{|\pi_i\rangle\}$, which we can also represent in terms of the rank-1 projection operators $\Pi_i = |\pi_i\rangle\langle\pi_i|$. These define a representation of all three-qubit states as probability vectors:

$$p(H_i) = \frac{1}{d}\text{tr}(\rho\Pi_i). \tag{4.6}$$

Given a probability distribution, we can construct the corresponding density matrix by the usual inversion formula, Eq. (2.6).

Note what happens if we take the expectation value of an operator:

$$\langle A \rangle = \text{tr}(A\rho). \tag{4.7}$$

Substituting in the expansion (2.6), we obtain

$$\langle A \rangle = \mathrm{tr}\left[A \sum_i \left((d+1)p(H_i) - \frac{1}{d} \right) \Pi_i \right] \tag{4.8}$$

$$= \sum_i \left((d+1)p(H_i) - \frac{1}{d} \right) \mathrm{tr}(A\Pi_i). \tag{4.9}$$

Denote the expectation value of an operator A given the SIC state Π_i as

$$\langle A : i \rangle = \mathrm{tr}(A\Pi_i). \tag{4.10}$$

Then,

$$\langle A \rangle = (d+1) \sum_i p(H_i)\langle A : i \rangle - \frac{1}{d} \sum_i \langle A : i \rangle. \tag{4.11}$$

We also know that

$$\sum_i \mathrm{tr}(A\Pi_i) = \mathrm{tr}\left[A \sum_i \Pi_i \right] = d\,\mathrm{tr}A. \tag{4.12}$$

Each of the four operators XXX, XYY, YXY and YYX are themselves traceless. If we fix

$$\mathrm{tr}A = 0, \tag{4.13}$$

then we obtain

$$\langle A \rangle = (d+1) \sum_i p(H_i)\langle A : i \rangle. \tag{4.14}$$

This applies to each of the four operators, and also to linear combinations of them. It is more appropriate to use it for the individual operators, since those correspond to individual experiments, or to single trials in a multi-trial experiment.

If we followed classical intuition, we might say, "The expectation value for the random variable A, if the system is in configuration Π_i, is some number $\langle A : i \rangle$. We don't know what configuration the system is really in, so we have some probability spread over i. To find the expectation value of A, we just weight the $\langle A : i \rangle$ according to those probabilities." However, this does not give the correct answer. The classical result is off by a factor $(d+1)$.

We can calculate the $\langle A : i \rangle$ for the Hoggar SIC. In fact, the peculiar symmetry of the Hoggar SIC makes the salient features of the computation rather easy to derive. The four operators in Mermin's inequality are elements of the group $\{D_k\}$ that generates the Hoggar SIC. Therefore, each of the four of them satisfies

$$|\langle \psi_i | D_k | \psi_i \rangle|^2 = \frac{1}{d+1} = \frac{1}{9}. \tag{4.15}$$

Furthermore, each operator D_k is Hermitian, so its eigenvalues are real, as is its expectation value given any state. Consequently,

$$\langle D_k : i \rangle = \langle \psi_i | D_k | \psi_i \rangle = \pm \frac{1}{3}. \tag{4.16}$$

This applies to each term in our linear combination of expectation values, Eq. (4.1). When we combine the expectation values for the four operators, the contributions might cancel each other, depending on the relative signs, but the absolute value of the sum total cannot exceed $4/3$. This is safely within the interval that a local hidden variable explanation could account for. So, the Hoggar SIC states cannot be used as to violate the three-qubit Bell inequality. This will remain true for any SIC that is generated from a fiducial by applying the three-qubit Pauli group.

By doing the algebra explicitly, we find that the Hoggar SIC states do not even reach the bound of $4/3$ that we deduced. In fact,

$$|\langle XXX : i \rangle - \langle XYY : i \rangle - \langle YXY : i \rangle - \langle YYX : i \rangle| = \frac{2}{3} \, \forall \, i. \tag{4.17}$$

Furthermore, any probabilistic combination of the Hoggar SIC states will also be consistent with the LHV bound. That is, if we pick a state from the Hoggar SIC following the probability distribution $p(H_i)$, then the linear combination of the four expectation values will stay safely in the classical region. If we then average over i, then this will remain true, no matter what the distribution $p(H_i)$ is.

However! The GHZ state itself corresponds to some probability distribution $p_{GHZ}(H_i)$, because we can write any state in the Hoggar SIC representation. Let the index $\mathcal{O}$ range over the four operators that we use to define the three-qubit Bell inequality:

$$\mathcal{O} \in \{XXX, -XYY, -YXY, -YYX\}. \tag{4.18}$$

For any of our four operators $\mathcal{O}$,

$$\sum_i p_{GHZ}(H_i)\langle \mathcal{O} : i \rangle = \frac{1}{9}, \tag{4.19}$$

meaning that the quantum expectation value is scaled up by the urgleichung's factor $d + 1$:

$$\langle \mathcal{O} \rangle = (d+1) \sum_i p_{GHZ}(H_i)\langle \mathcal{O} : i \rangle = 1. \tag{4.20}$$

This is just restating the fact that the GHZ state is a simultaneous eigenstate of the four operators. Therefore,

$$\sum_{\mathcal{O}} \sum_i p_{GHZ}(H_i)\langle \mathcal{O} : i \rangle = \frac{4}{9}. \tag{4.21}$$

This value lies within the classical interval $[-2, 2]$, but when we account for the extra factor in the urgleichung, we find

$$(d + 1) \sum_{\mathcal{O}} \sum_{i} p_{\text{GHZ}}(H_i) \langle \mathcal{O} : i \rangle = 4. \tag{4.22}$$

It is that factor of $(d + 1)$ that lifts us over the edge into nonclassical territory.

One way to interpret this result is as a bridge between *interference experiments* and *Bell–Kochen–Specker phenomena*. Interference phenomena are weakly nonclassical: That is, the *bare fact* of interference can occur in fundamentally classical theories [6, 32]. However, by adopting the proper mindset, we can strengthen the double-slit experiment into a genuine test for nonclassicality.

Interference between *nonorthogonal alternatives*—in other words, between alternative paths represented by nonorthogonal quantum states—can be a stronger test of nonclassicality than the double-slit experiment as it is normally described. This is because generalizing to nonorthogonal states allows the "which-way" information to be the outcome of an informationally complete measurement. (Heuristically speaking, this ties in with the idea that pre- and post-selection effects with nonorthogonal states are more strongly nonclassical than they are when one considers only orthogonal states [32, 33].).

Mermin wrote that the n-qubit GHZ state "combines two of the most peculiar features of the quantum theory" [9], interference of probabilities and the failure of local hidden-variable explanations. Using the Hoggar SIC, we have found a concise expression of this when $n = 3$. Correlations that violate the three-qubit Bell inequality encode a kind of interference that defies mimicking by classical randomness.

Mermin's three-qubit Bell inequality is closely related to the GHZ thought-experiment, which is sometimes touted as an example where the distinction between quantum and classical is "all-or-nothing". The hypothesis of local, intrinsic hidden variables implies one result with certainty, and quantum mechanics implies another, also with certainty. Stated carelessly, this can create the impression that probabilities are not involved. But a prediction made with probability 0 or 1 is still a probabilistic statement. Moreover, we can see the nontrivial probabilities churning just below the surface.

In the GHZ scenario, Alice measures the X observable on each of her three qubits and checks the parity of the answer. Writing $|+\rangle$ and $|-\rangle$ for the eigenstates of X, and denoting the SIC representation of the state $|+ + +\rangle$ by p_{+++}, she calculates that

$$\sum_{i} p_{\text{GHZ}}(H_i) p_{+++}(H_i) = \frac{5}{288}. \tag{4.23}$$

The same result holds for the other states of the same parity, p_{+--}, p_{-+-} and p_{--+}. Classical intuition would lead her to say that this number is the probability for obtaining each of the odd-parity outcomes, given a preparation described by p_{GHZ}. In turn, the probability for getting *any* odd-parity outcome would be the sum of the probabilities for the four alternatives. But she knows to take the quantum correction, which is given by the urgleichung:

$$P(\text{odd}) = d(d+1) \sum_i p_{\text{GHZ}}(H_i)[p_{+++}(H_i) + p_{+--}(H_i) + p_{-+-}(H_i) + p_{--+}(H_i)]$$

$$- \sum_i [p_{+++}(H_i) + p_{+--}(H_i) + p_{-+-}(H_i) + p_{--+}(H_i)]. \tag{4.24}$$

This evaluates to

$$P(\text{odd}) = 72 \cdot 4 \cdot \frac{5}{288} - 4 = 1. \tag{4.25}$$

So, while ascribing the GHZ state does imply predictions with probability unity, that unity arises from the combination of many fractions.

4.3 Qubit Pairs and Twinned Tetrahedral SICs

In this section, we change perspective slightly. Instead of applying one SIC measurement to the entirety of a tripartite system, we start with a smaller SIC and apply measurements based on it to each of two halves of a bipartite system. The end result will be a sharpened intuition for the nonclassicality of qubit pairs.

We have seen how attempting to interpret a SIC outcome as a specific, pre-existing physical property leads to a contradiction with the predictions of quantum theory. Any assumption which would incline us to interpret SIC outcomes in this way is, therefore, an assumption that would lead the unwitting physicist into error and would stand in the way of using quantum theory fruitfully. We can identify one such counterproductive idea—the *EPR criterion of reality* [34]:

> If, without in any way disturbing a system one can [gather the information required to] predict with certainty (i.e., with probability equal to unity) the value of a physical quantity, then there exists an element of physical reality corresponding to this physical quantity.

We now present a scenario in which the EPR criterion leads the unwitting physicist to conclude that SIC outcomes are pre-existing, specific "elements of physical reality."

Alice arranges the following experiment. A device produces pairs of qubits to which Alice ascribes a maximally entangled state. Each qubit then travels to one of two widely separated instruments, which we can designate the left detector and the right detector. The detectors each have a control knob that can be turned to four different settings. Alice models the detectors using binary POVMs defined using the states comprising two SICs. The first SIC, which we can denote $\{\Pi_i^+\}$, is a set of four projectors that together form the vertices of a tetrahedron inscribed in the Bloch sphere. The second SIC, $\{\Pi_i^-\}$, forms the tetrahedron whose vertices are antipodal to those of the first. Together, the two tetrahedra form a stellated octahedron. When the knob on a detector is set to position i, it implements the POVM

$$\{\Pi_i^+, I - \Pi_i^+\} = \{\Pi_i^+, \Pi_i^-\}. \tag{4.26}$$

Consider first the case when Alice sets the two control knobs to the same position. She performs the measurement with one detector, say the one on the left. If she experiences the $+$ outcome, she can predict with 100% certainty that she would experience the $-$ outcome, if she were to walk over to the right-hand detector and test the other qubit. Likewise, if she experiences the $-$ outcome on the left, she can predict with a probability of unity that she will experience $+$ upon using the detector on the right. This holds true for all four values of the control setting i.

Alice, deciding to entertain the EPR criterion, concludes that there exists within both particles emitted from the common source an "element of physical reality" that implies the outcome of each of the binary tests.

What happens when Alice chooses to set the two detectors differently? Now, if she performs test i on the left and obtains the $+$ outcome, she updates her state for the right-hand particle to Π_i^-. She then performs the test for some detector setting $j \neq i$ on the right. Her probability of obtaining the $-$ outcome on the right is

$$\mathrm{tr}(\Pi_i^- \Pi_j^-) = \frac{1}{d+1} = \frac{1}{3}. \tag{4.27}$$

Likewise, if Alice first experiences the $-$ outcome on the left, she updates her state for the right-hand side to Π_i^+, and her probability for obtaining the $+$ result on the right is

$$\mathrm{tr}(\Pi_i^+ \Pi_j^+) = \frac{1}{3}. \tag{4.28}$$

In summary, when Alice sets the detector controls to the same position, her probability of an anti-coincidence ($+$ on one device, $-$ on the other) is unity. If she sets the detector controls differently, her probability of anti-coincidence is 1/3.

Can Alice account for these results in terms of hidden variables? Guided by the EPR criterion, she postulates that each particle carries an "instruction set" [35] of the form $\lambda_0 \lambda_1 \lambda_2 \lambda_3$. Each λ_i is a pre-existing physical property of some kind, intrinsic to a particle, which can be thought of as taking values in the set $\{+, -\}$. The value of λ_i specifies the outcome of testing that particle with a detector configured to setting i.

Alice hypothesizes that the source produces particle pairs with anticorrelated instruction sets:

$$\left\{ \begin{array}{l} + + - - \\ - - + + \\ - + + - \\ + - - + \\ + - + - \\ - + - + \end{array} \right\} \text{ with } \left\{ \begin{array}{l} - - + + \\ + + - - \\ + - - + \\ - + + - \\ - + - + \\ + - + - \end{array} \right\}. \tag{4.29}$$

Alice finds that whichever instruction sets the particles carry, if she configures her two detectors identically, these instruction sets imply perfect anti-coincidence. If she instead sets her detector knobs to different positions, each choice of detector configurations will produce anti-coincidence with probability $1/3$, provided that all six of these instruction-set pairs occur with equal probability.

How should Alice proceed from this point? She supposes, as a physicist naturally would, that whatever an instruction set is, a particle can carry one all by itself. The source in this experiment, Alice figures, happens to produce particles in pairs with perfectly anti-correlated instruction sets. To imagine that a particle *only* has an instruction set when it is produced as half of a pair strikes her as a touch pathological. A spinning top has angular momentum whether or not it is started into motion at the same time as another top, spun in the opposite direction.

Let $T(i)$ be Alice's probability for obtaining the $+$ outcome when performing test i on an isolated system. Quantum theory tells us that we can craft another measurement corresponding to the four-outcome POVM

$$\left\{ \frac{1}{2}\Pi_0, \frac{1}{2}\Pi_1, \frac{1}{2}\Pi_2, \frac{1}{2}\Pi_3 \right\}. \tag{4.30}$$

Alice's probability for obtaining outcome i in this experiment is

$$p(R_i) = \frac{1}{d}\mathrm{tr}(\rho\Pi_i) = \frac{1}{2}T(i). \tag{4.31}$$

What is Alice's interpretation of this four-outcome experiment in terms of her hidden-variable hypothesis? Suppose that she has $T(0) = 1$. In quantum language, this means that her state for the system is the projector Π_0^+. Referring back to the instruction sets listed in Eq. (4.29), Alice notes that three of them predict $+$ for the binary test on the first element:

$$\{++--, +--+, +-+-\}. \tag{4.32}$$

Selecting a $+$ at random from this list, Alice finds that she obtains a $+$ in position 0 with probability $1/2$, and in each of the other positions with probability $1/6$. So, she can interpret $p(R_i)$ as the probability that a $+$ sign, chosen at random from all the $+$ signs occurring in all possible instruction sets, falls in position i.

The hypothesis of instruction sets implies that the outcome of a tetrahedral SIC measurement, Eq. (4.30), is a classical random variable. To adapt Einstein's phrase, the SIC outcome is there even when nobody looks. Knowing that the SIC measurement is informationally complete, and seeing that its outcome probabilities are determined by the probability distribution over the six instruction sets, we conclude that the distribution over the instruction sets is all that is necessary to calculate the outcome statistics for *any* experiment.

There is another route to the instruction sets in Eq. (4.29), which begins with a set of desiderata that Spekkens provides for a hidden-variable model [36]. The guiding philosophy of the Spekkens criteria is that two quantities which imply the same statistics should have the same representation in terms of probability distributions over the underlying hidden variables. If two preparations of a system yield the same statistics for all possible measurements, then those two preparations correspond to the same distribution over λ. Likewise, if two measurements have the same statistics for all possible preparations, then those two measurements correspond to the same conditional probabilities of outcomes given λ's. The key quantities are effects, that is, positive semidefinite operators that satisfy

$$0 < E \leq I. \tag{4.33}$$

Call the set of all effects $\mathcal{E}$. By hypothesis, if Alice knows the underlying "ontic state" λ, she has a map from effects to probabilities:

$$w : \mathcal{E} \to [0, 1]. \tag{4.34}$$

The function w will generally depend upon λ. What properties will it satisfy? First, it obeys a sum rule. For any discrete set of effects $\{E_i\} \subset \mathcal{E}$, if $\sum_i E_i$ is also an effect, then

$$w\left(\sum_i E_i\right) = \sum_i w(E_i). \tag{4.35}$$

We will only need the particular special case of this in which the sum of the $\{E_i\}$ is the identity operator, i.e., when the set of effects is a POVM. This is equivalent to saying that whatever the underlying ontic state of the system, when Alice applies a measurement, she is sure that *something* has to happen.

Furthermore, for any effect $E \in \mathcal{E}$ and real number $s \in [0, 1]$, if $sE \in \mathcal{E}$, then

$$w(sE) = sw(E). \tag{4.36}$$

Again, we will only need a special case of this, specifically the case when $s = 1/2$. This is equivalent to saying that for any measurement, we can post-process the outcome by flipping a fair coin.

The identity effect is assigned unit probability:

$$w(I) = 1. \tag{4.37}$$

If Alice doesn't care at all about what she does, then her probability of "whatever" happening is 1, regardless of the ontic state. When else can she have certainty? If and only if the effect in question is a projection [37]:

$$w(E) \in \{0, 1\} \text{ if and only if } E^2 = E. \tag{4.38}$$

What do these conditions imply for a qubit SIC? First, the SIC states form a POVM when scaled down by the dimension:

$$\sum_i \frac{1}{2}\Pi_i = I. \tag{4.39}$$

Therefore, it must be the case that

$$\sum_i w\left(\frac{1}{2}\Pi_i\right) = 1. \tag{4.40}$$

In turn, by the post-processing assumption,

$$\sum_i w\left(\frac{1}{2}\Pi_i\right) = \frac{1}{2}\sum_i w(\Pi_i). \tag{4.41}$$

Each Π_i is a projector, so each $w(\Pi_i)$ on the right-hand side must be either 0 or 1. Because the sum total must be normalized, exactly two terms are 0, while the other two both equal 1. Consequently, the instruction sets in Eq. (4.29) are the only configurations of hidden variables that are compatible with our basic desiderata and with the structure of a qubit SIC measurement.

If we postulate that a tetrahedral SIC measurement $\{\frac{1}{2}\Pi_i^+\}$ is possible, and we assert that the hidden-variable description of the qubit meets the Spekkensian standards, then any quantum state for the qubit implies a probability distribution $\varrho(\lambda)$ over the six instruction sets in Eq. (4.29). In turn, such a probability distribution implies a $p(R_i)$, specified by

$$p(R_i) = \frac{1}{2}\sum_\lambda \varrho(\lambda)\delta_{\lambda_i,+}. \tag{4.42}$$

This has a ready interpretation in terms of a two-step stochastic process. Effectively, we are picking an instruction set at random with probability $\varrho(\lambda)$, and then we are flipping a fair coin to select one of the two + signs in that instruction set.

By mapping points in the Bloch ball to density operators, and then solving for the corresponding hidden-variable distributions, we can map out the "classical region" of qubit state space. We define this region to be the subset of state space within which all elements of ϱ turn out nonnegative, meaning that the vector ϱ can be interpreted as an ordinary probability distribution, rather than a quasi-probability one. The eight states that comprise the vertices of the SICs $\{\Pi_i^+\}$ and $\{\Pi_i^-\}$ are classical, by this standard. The classical region of state space is the cube that is their convex hull.

Each of the six instruction sets in Eq. (4.29) is a "dispersion-free state" [24–27]. Using the SIC representation of qubit state space, we can see that they do not correspond to valid quantum states. Each instruction set implies a probability distribution p in which two elements equal $1/2$ and the other two equal 0. Using the inversion formula, we can map these probability distributions to linear operators. The resulting operators will all be Hermitian, but they *will not* be positive semidefinite. Therefore, the dispersion-free states cannot be quantum states. Pictorially, they can be represented as the vertices of an octahedron *outside* the Bloch sphere: While the Bloch sphere has radius 1, the dispersion-free states all reside at a distance of $\sqrt{3}$ from the origin.

We have seen that if we try to model the SIC states as essentially classical, then the eigenstates of the Pauli operators become maximally quantum, in that they lie as far as possible from the region of the Bloch ball for which a classical model exists. This is in a sense the dual of the statement that qubit SIC states are "magic states" for quantum computation when the Pauli eigenstates are treated as classical [38]. Consequently, we now have a certain intuition for the result of Andersson et al., who find that the maximal violation in an "elegant" two-qubit Bell inequality occurs when the measurements on one qubit are the Pauli eigenbases and the measurements on the other are the binary tests defined by pairs of antipodal SIC vectors [39]. Recent work in this vein has made a connection between SICs, Bell inequalities and quantum random number generation [40].

4.4   Failure of Hidden Variables for Qutrits

The Spekkens criteria for hidden-variable models provide an alternative perspective on a Kochen–Specker proof that Bengtsson, Blanchfield and Cabello derive for qutrit systems [41]. Their proof relies upon a set of 21 vectors, 9 of which comprise a SIC and the other 12 of which form a particular set of orthonormal bases. These four bases have the nice property that they are all *mutually unbiased* with respect to one another. That is, the overlap of any vector from one basis with any vector from another basis is constant. In the Bengtsson et al. construction, the only properties that matter are the orthogonalities among the 21 vectors; by employing the fact that the set of 9 specifically form an informationally complete POVM, we can appreciate the result in a new way.

First, we note that if we have three vectors that form an orthonormal basis for $\mathbb{C}^3$, then the projectors onto those vectors add to the identity operator, meaning that by the Spekkens rules,

$$w(E_1) + w(E_2) + w(E_3) = 1. \tag{4.43}$$

Furthermore, each term in the sum must be either 0 or 1, implying that whatever the underlying ontic state, exactly one vector in any orthonormal basis is assigned probability 1.

Now, we consider the cat's cradle of vectors we encounter in dimension $d = 3$. First, there's the Hesse SIC. Take $\omega = e^{2\pi i/3}$, and construct the set of states $\{|\pi_j\rangle\}$ given by the columns of

$$\frac{1}{\sqrt{2}} \begin{pmatrix} 0 & -1 & 1 & 0 & -1 & 1 & 0 & -1 & 1 \\ 1 & 0 & -1 & \omega & 0 & -\omega & \omega^2 & 0 & -\omega^2 \\ -1 & 1 & 0 & -\omega^2 & \omega^2 & 0 & -\omega & \omega & 0 \end{pmatrix}. \tag{4.44}$$

We have a duality relation between the canonical mutually unbiased bases and the Hesse SIC. We recall from Chap. 3 that this relation is rather intricate: Each of the 9 SIC states is orthogonal to exactly 4 of the MUB states, and each of the MUB states is orthogonal to exactly 3 SIC states [19].

Because Π_1 is orthogonal to the MUB element M_{123}, if the underlying ontic state λ implies $w(M_{123}) = 1$, then $w(\Pi_1) = 0$. The more of the $\{w(\Pi_i)\}$ that we can "zero out" in this way, the smaller their sum will be. By working through all the possibilities for assigning a w of unity to exactly one element of each basis, it is straightforward to show that whatever λ might be,

$$\sum_i w(\Pi_i) \leq 2. \tag{4.45}$$

But from the post-processing rule,

$$w\left(\frac{1}{d}\Pi_i\right) = \frac{1}{d}w(\Pi_i), \tag{4.46}$$

and from the sum rule,

$$\sum_i w\left(\frac{1}{d}\Pi_i\right) = w\left(\sum_i \frac{1}{d}\Pi_i\right) = w(I) = 1. \tag{4.47}$$

Therefore,

$$\sum_i w(\Pi_i) = d \sum_i w\left(\frac{1}{d}\Pi_i\right) = 3. \tag{4.48}$$

Our plans for a hidden-variable model have gone awry. The SIC states burst out of the confines that the orthonormal bases establish. The set of 9 and the set of 12 cannot coexist in the world of λ: If we take one set to have a classical representation, then the other cannot.

Bengtsson et al. derive a contradiction between the hypothesis of intrinsic hidden variables and the predictions of quantum theory by invoking the Born rule. This is equivalent to postulating a probability assignment for all POVMs. From this, the POVM version of Gleason's theorem establishes that the state space is the space of unit-trace positive semidefinite operators and that probabilities are calculated by the inner product between density matrices and effects [42, 43]. Accordingly, we can say that

$$\sum_i \langle \Pi_i \rangle = \sum_i \text{tr}(\rho \Pi_i) = d \, \text{tr}\rho = 3. \tag{4.49}$$

By invoking Spekkens' criteria for a hidden-variable model, we see that we do not have to postulate outcome statistics for all POVMs. Instead, we can derive the desired contradiction by considering only a discrete set of rays in the Hilbert space $\mathbb{C}^3$.

4.5 Quantum Theory from Nonclassical Probability Meshing

Let us now return to the Hoggar SIC. We have seen how this configuration, and the representation of quantum state space that it furnishes, provides a link between *interference phenomena* and the *failure of hidden-variable models*. This mathematical construction—just sixty-four complex lines, making equal angles with one another—evidently cuts quite deeply into the quantum mysteries. Consider again our expression for the expectation value of an operator:

$$\langle A \rangle = (d+1) \sum_i p(R_i)\langle A : i \rangle - \frac{1}{d} \sum_i \langle A : i \rangle. \qquad (4.50)$$

Seen in one way, this formula is a way to upstage the textbook double-slit experiment: Like the double slit, our "d^2-nonorthogonal-slit experiment" captures the counterintuitive way quantum theory requires us to use expectations for one scenario to make deductions about another. However, it indicates a kind of interference that cannot be emulated by classical stochasticity. And, seen in another way, it opens the possibility of violating a Bell inequality.

This is a sufficiently appealing notion that one is naturally tempted to wonder how far it can go. If we take this idea as basic, if we make this way of relating expectations between counterfactual scenarios as a fundamental precept, what can we derive from it? The answer, potentially, is *quantum theory itself.*

During the twenty-first century, there has been increasing interest in the project of *rederiving* or *reconstructing* the mathematical apparatus of quantum mechanics, by starting with a set of basic principles that, one hopes, are more meaningful or illuminating than the abstruse invocations with which one traditionally begins the subject [20]. These efforts begin with a set of axioms, typically expressed in operationalist terms as statements about what kinds of laboratory procedures are possible, and rederive quantum theory from that starting point [44–50]. Mathematically, these derivations are successful; however, they share the common feature that they make quantum theory as "benignly humdrum" as possible [20]. The remarkable and enigmatic phenomena seen within quantum physics are no closer to the surface than they were in the standard presentation of the formalism [29]. Indeed, the notions invoked in these axioms are often not that quantum at all. For example, the system of Chiribella, D'Ariano and Perinotti relies upon the purifiability of mixed states [51], which was originally discovered in quantum physics but actually arises naturally in the Spekkens toy model, a fundamentally classical theory [6, 52]. Moreover, it is not at first glance clear what these proposed sets of axioms have in common with each other, other than their conclusion.

In contrast, one research program aims to take the urgleichung as a basic postulate upon which quantum theory can be built [18, 20, 29, 53, 54]. The urgleichung embodies a rejection of the hypothesis of hidden variables, phrased in a way that does not depend upon the ordinary textbook formalism of Hilbert spaces and operators.

The hope is that whatever deep lesson quantum physics has to teach us about the character of the natural world, we will see it most clearly by bringing the essential expression of it to the forefront, rather than deriving it as a consequence of "benignly humdrum" axioms. Postulating that the urgleichung is fundamental—that, try as one might to establish a standard reference measurement, nonclassical "deformation" of probabilities *cannot be avoided*—foregrounds the strangeness of quantum theory. What we know so far is that a reconstruction on these lines *can* be done, but at the price of invoking a couple additional presumptions that seem too specific to belong in the final answer. My own suspicion is that these additional requirements are stronger and more particular than is truly necessary. This is where drawing upon a variety of other reconstruction efforts may be helpful: They suggest that certain technical matters arising in the course of a reconstruction (e.g., the choice of a particular symmetry group) can be dispatched with relative ease.

In special relativity and in thermodynamics, one builds up the theory starting from postulates, the first of which has the character of a guarantee. (Inertial observers Alice and Bob can come to agree on the laws of physics; energy is conserved.) The second is a foil to the first, frustrating it and generating a degree of dramatic tension. (Alice and Bob cannot agree on a standard of rest, even by measuring the speed of light; entropy is nondecreasing.) Then comes a statement of unattainability, which is derived in one case—massive bodies cannot attain light speed—and assumed in the other—we cannot cool all the way to absolute zero. We might also draw an analogy between the clock postulate of special relativity, which lets us analyze accelerated motion using momentarily co-moving inertial frames, and the zeroth law of thermodynamics, which as it is applied in practice is a statement about momentary equilibrium between systems being a transitive condition. Might a similar story hold true for quantum mechanics as well? In the view of quantum theory we are developing here, the possibility of probability-1 predictions might be considered a guarantee: *Certainty is allowed.* The urgleichung is then an axiom of frustration: *Certainty cannot be about hidden variables.* Perhaps the rejection of hidden variables, carefully formulated in the urgleichung, will one day be recognized as the Second Law of Quantum Mechanics.

References

1. E. Schrödinger, Discussion of probability relations between separated subsystems. Math. Proc. Cam. Phil. Soc. **31**(4), 555–63 (1935). https://doi.org/10.1017/S0305004100013554
2. J. Barrett, Information processing in generalized probabilistic theories. Phys. Rev. A **75**(3), 032304 (2007). https://doi.org/10.1103/PhysRevA.75.032304
3. H. Barnum, J. Barrett, M. Leifer, A. Wilce, Teleportation in general probabilistic theories. Proc. Symp. Appl. Math. **71**, 25–48 (2012)
4. R.F. Werner, Quantum states with Einstein-Podolsky-Rosen correlations admitting a hidden-variable model. Phys. Rev. A **40**(6), 4277 (1989). https://doi.org/10.1103/PhysRevA.40.4277
5. D. Gottesman, The Heisenberg representation of quantum computers, in *Group22: Proceedings of the XXII International Colloquium on Group Theoretical Methods in Physics*, eds. by S.P. Corney, R. Delbourgo, P.D. Jarvis (International Press, Vienna, 1999)

6. R.W. Spekkens, Evidence for the epistemic view of quantum states: a toy theory. Phys. Rev. A **75**(3), 032110 (2007). https://doi.org/10.1103/PhysRevA.75.032110

7. R.W. Spekkens, Quasi-quantization: classical statistical theories with an epistemic restriction, in *Quantum Theory: Informational Foundations and Foils*, eds. by G. Chiribella, R.W. Spekkens (Springer, Berlin, 2016), pp. 83–135. https://doi.org/10.1007/978-94-017-7303-4_4

8. N.D. Mermin, What's wrong with these elements of reality? Phys. Today **43**(6), 9 (1990). https://doi.org/10.1063/1.2810588; Reprinted in Why Quark Rhymes With Pork (Cambridge University Press, Cambridge, 2016), pp. 43–49

9. N.D. Mermin, Extreme quantum entanglement in a superposition of macroscopically distinct states. Phys. Rev. Lett. **65**(15), 1838–40 (1990). https://doi.org/10.1103/PhysRevLett.65.1838

10. N.D. Mermin, Hidden variables and the two theorems of John Bell. Rev. Mod. Phys. **65**(3), 803–15 (1993). https://doi.org/10.1103/RevModPhys.65.803

11. N.D. Mermin, Erratum: hidden variables and the two theorems of John Bell. Rev. Mod. Phys. **88**(3), 039902 (2016). https://doi.org/10.1103/RevModPhys.88.039902

12. S.G. Hoggar, Two quaternionic 4-polytopes, in *The Geometric Vein: The Coxeter Festschrift*, eds. by C. Davis, B. Grünbaum, F.A. Sherk (Springer, Berlin, 1981). https://doi.org/10.1007/978-1-4612-5648-9_14

13. S.G. Hoggar, 64 lines from a quaternionic polytope. Geom. Dedicata. **69**, 287–289 (1998). https://doi.org/10.1023/A:1005009727232

14. A. Szymusiak, W. Słomczyński, Informational power of the Hoggar symmetric informationally complete positive operator-valued measure. Phys. Rev. A **94**, 012122 (2015). https://doi.org/10.1103/PhysRevA.94.012122

15. B.C. Stacey, Sporadic SICs and the normed division algebras. Found. Phys. **47**, 1060–64 (2017). https://doi.org/10.1007/s10701-017-0087-2

16. B.C. Stacey, Geometric and information-theoretic properties of the Hoggar lines (2016). arXiv:1609.03075

17. W.K. Wootters, Symmetric informationally complete measurements: can we make big ones out of small ones? Video (2009). http://pirsa.org/09120023/

18. C.A. Fuchs, R. Schack, Quantum-Bayesian coherence. Rev. Mod. Phys. **85**, 1693–1715 (2013). https://doi.org/10.1103/RevModPhys.85.1693

19. B.C. Stacey, SIC-POVMs and compatibility among quantum states. Mathematics **4**(2), 36 (2016). https://doi.org/10.3390/math4020036

20. C.A. Fuchs, B.C. Stacey, Some negative remarks on operational approaches to quantum theory, in *Quantum Theory: Informational Foundations and Foils*, eds. by G. Chiribella, R.W. Spekkens (Springer, Berlin, 2016), pp. 283–305. https://doi.org/10.1007/978-94-017-7303-4_9

21. C.A. Fuchs, B.C. Stacey, QBism: quantum theory as a hero's handbook (2016). arXiv:1612.07308

22. C.A. Fuchs, Notwithstanding Bohr, the reasons for QBism. Mind Matter **15**(2), 245–300 (2017)

23. J.B. DeBrota, C.A. Fuchs, B.C. Stacey, Symmetric informationally complete measurements identify the irreducible difference between classical and quantum systems. Phys. Rev. Res. **2**, 013074 (2020). https://doi.org/10.1103/PhysRevResearch.2.013074

24. J. von Neumann, *Mathematical Foundations of Quantum Mechanics* (Princeton University Press, Princeton, 1955)

25. R.F. Werner, Comment on Maudlin's paper 'What Bell did'. J. Phys. A **47**(42), 424011 (2014). https://doi.org/10.1088/1751-8113/47/42/424011

26. B.C. Stacey, Von Neumann was not a quantum Bayesian. Phil. Trans. Roy. Soc. A **374**, 20150235 (2016). https://doi.org/10.1098/rsta.2015.0235

27. N.D. Mermin, R. Schack, Homer nodded: von Neumann's surprising oversight. Found. Phys. **48**, 1007–20 (2018). https://doi.org/10.1007/s10701-018-0197-5

28. C.A. Fuchs, N.D. Mermin, R. Schack, An introduction to QBism with an application to the locality of quantum mechanics. Am. J. Phys. **82**(8), 749–54 (2014). https://doi.org/10.1119/1.4874855

29. M. Appleby, C.A. Fuchs, B.C. Stacey, H. Zhu, Introducing the Qplex: a novel arena for quantum theory. Euro. Phys. J. D **71**, 197 (2017). https://doi.org/10.1140/epjd/e2017-80024-y

30. J.M. Renes, R. Blume-Kohout, A.J. Scott, C.M. Caves, Symmetric informationally complete quantum measurements. J. Math. Phys. **45**, 2171–2180 (2004). https://doi.org/10.1063/1.1737053
31. J. Jedwab, A. Wiebe, A simple construction of complex equiangular lines, in *Algebraic Design Theory and Hadamard Matrices* (Springer, Berlin, 2015), pp. 159–169. https://doi.org/10.1007/978-3-319-17729-8_13
32. R.W. Spekkens, Reassessing claims of nonclassicality for quantum interference phenomena (2016). http://pirsa.org/16060102/
33. M.F. Pusey, M.S. Leifer, Logical pre- and post-selection paradoxes are proofs of contextuality. EPTCS **195**, 295–306 (2015). https://doi.org/10.4204/EPTCS.195.22
34. A. Einstein, B. Podolsky, N. Rosen, Can quantum-mechanical description of physical reality be considered complete? Phys. Rev. **47**, 777–80 (1935). https://doi.org/10.1103/PhysRev.47.777
35. N.D. Mermin, Is the moon there when nobody looks? Reality and the quantum theory. Phys. Today **38**(4), 38–47 (1985). https://doi.org/10.1063/1.880968
36. R.W. Spekkens, The status of determinism in proofs of the impossibility of a noncontextual model of quantum theory. Found. Phys. **44**(11), 1125–55 (2014). https://doi.org/10.1007/s10701-014-9833-x
37. R.W. Spekkens, Contextuality for preparations, transformations, and unsharp measurements. Phys. Rev. A **71**(5), 052108 (2005). https://doi.org/10.1103/PhysRevA.71.052108
38. S. Bravyi, A. Kitaev, Universal quantum computation with ideal Clifford gates and noisy ancillas. Phys. Rev. A **71**, 022316 (2005). https://doi.org/10.1103/PhysRevA.71.022316
39. O. Andersson, P. Badzląg, I. Bengtsson, I. Dumitru, A. Cabello, Self-testing properties of Gisin's elegant Bell inequality. Phys. Rev. A **96**, 032119 (2017). https://doi.org/10.1103/PhysRevA.96.032119
40. A. Tavakoli, M. Farkas, D. Rosset, J.D. Bancal, J. Kaniewski, Mutually unbiased bases and symmetric informationally complete measurements in bell experiments. Sci. Adv. **7**, eabc3847 (2021). https://doi.org/10.1126/sciadv.abc3847
41. I. Bengtsson, K. Blanchfield, A. Cabello, A Kochen-Specker inequality from a SIC. Phys. Lett. A **376**, 374–376 (2012). https://doi.org/10.1016/j.physleta.2011.12.011
42. P. Busch, Quantum states and generalized observables: a simple proof of Gleason's theorem. Phys. Rev. Lett. **91**, 120403 (2003). https://doi.org/10.1103/PhysRevLett.91.120403
43. C.M. Caves, C.A. Fuchs, K.K. Manne, J.M. Renes, Gleason-type derivations of the quantum probability rule for generalized measurements. Found. Phys. **34**, 193–209 (2004). https://doi.org/10.1023/B:FOOP.0000019581.00318.a5
44. H. Barnum, M.P. Müller, C. Ududec, Higher-order interference and single-system postulates characterizing quantum theory. New J. Phys. **16**, 123029 (2014). https://doi.org/10.1088/1367-2630/16/12/123029
45. H. Barnum, M. Graydon, A. Wilce, Some nearly quantum theories. EPTCS **195**, 59–70 (2015)
46. A. Wilce, A royal road to quantum theory (or thereabouts) (2016). arXiv:1606.09306
47. M. Krumm, H. Barnum, J. Barrett, M.P. Müller, Thermodynamics and the structure of quantum theory (2016). arXiv:1608.04461
48. A. Cabello, A simple explanation of Born's rule (2018). arXiv:1801.06347
49. J. van de Wetering, Sequential measurement characterizes quantum theory (2018). arXiv:1803.11139
50. H. Barnum, J. Hilgert, Strongly symmetric spectral convex bodies are Jordan state spaces (2019). arXiv:1904.03753
51. G. Chiribella, G.M. D'Ariano, P. Perinotti, Informational derivation of quantum theory. Phys. Rev. A **84**(1), 012311 (2011). https://doi.org/10.1103/PhysRevA.84.012311
52. L. Disilvestro, D. Markham, Quantum protocols within Spekkens' toy model. Phys. Rev. A **95**(5), 052324 (2017). https://doi.org/10.1103/PhysRevA.95.052324
53. B.C. Stacey, Quantum theory as symmetry broken by vitality (2019). arXiv:1907.02432
54. J.B. DeBrota, C.A. Fuchs, J.L. Pienaar, B.C. Stacey, The Born rule as Dutch-book coherence (and only a little more) (2020). arXiv:2012.14397

Chapter 5
The Hoggar-Type SICs

We take a tour of a particularly special set of equiangular lines in eight-dimensional complex Hilbert space. In one language, this structure is an equiangular tight frame. In the jargon of another field, it defines an informationally complete measurement, that is, a way to represent all quantum states of three-qubit systems as probability distributions. Investigating the shape of this representation of state space yields a pattern of connections among a remarkable spread of mathematical constructions. In particular, studying the Shannon entropy of probabilistic representations of quantum states leads to an intriguing link between the questions of real and of complex equiangular lines. Furthermore, we will find relations between quantum information theory and mathematical topics like octonionic integers and the 28 bitangents to a quartic curve.

5.1 Introduction

When we study quantum physics, or electrical engineering, or perhaps even high-school algebra, we learn that we can double the dimension of the real number line to get the complex plane. A convenient way to describe this doubling is to say we take ordered pairs of real numbers (a, b) and declare that they add entrywise and multiply as $(ac - bd, ad + bc)$. Nothing stops us from repeating this process, making ordered pairs of complex numbers and establishing their arithmetic. The most useful way to do so introduces a conjugation into the game:

$$(a, b)(c, d) := (ac - db^*, a^*d + cb), \tag{5.1}$$

which reduces to the previous rule if every number is its own conjugate, as in $\mathbb{R}$. The conjugate of an ordered pair of complex numbers is

© The Author(s), under exclusive license to Springer Nature Switzerland AG 2021   57
B. C. Stacey, *A First Course in the Sporadic SICs*,
SpringerBriefs in Mathematical Physics 41,
https://doi.org/10.1007/978-3-030-76104-2_5

$$(a, b)^* := (a^*, -b), \tag{5.2}$$

which also reduces to the familiar complex conjugation when a and b fall on the real line. With these definitions, ordered pairs of complex numbers become *quaternions,* the set of which we write $\mathbb{H}$ for William Rowan Hamilton.

Repeating the doubling process again yields the *octonions* $\mathbb{O}$. Each octonion is specified by a list of eight real numbers. Where the complex numbers have i, the quaternions have i, j and k; and the octonions have i_1 through i_7. In an important sense, the doubling process stops here. With each step, the number system we construct loses a familiar property: In $\mathbb{H}$ multiplication is not commutative, in $\mathbb{O}$ it is not even associative, and if we try to double again, we won't even have an absolute value that works properly. The technical statement is *Hurwitz's theorem,* which says that if we want $|xy| = |x||y|$ to hold, and our original field is the real numbers, then the only possibilities are $\mathbb{R}$, $\mathbb{C}$, $\mathbb{H}$ and $\mathbb{O}$.

Much of this has percolated out into the pop-math world. A little less familiar is the question of how we might find *integers* within these higher-dimensional number systems. Within the real line, of course, we know the ordinary integers $\mathbb{Z}$, which we can make by adding up $+1$ and -1. In the complex plane, there are actually two different options, each comprising a regularly spaced pattern of points. The *Gaussian* integers are the square lattice made by taking all the complex numbers whose real and imaginary parts both belong to $\mathbb{Z}$. The elements of absolute value 1 are, in familiar notation, ± 1 and $\pm i$. Meanwhile, the *Eisenstein* integers are a triangular lattice, given by $a + b\omega$ with $a, b \in \mathbb{Z}$ and $\omega := e^{2\pi i/3}$. In the Eisenstein integers, the unit-norm elements are the vertice of a regular hexagon: ± 1, $\pm\omega$ and $\pm\omega^2$.

There are also two options for a lattice of integers in the quaternions. The *Hurwitz integers* are the quaternions whose coefficients are either all integers in $\mathbb{Z}$ or all half-integers. The other option, the Lipschitz integers, are the subring of the Hurwitz integers whose coefficients are all integral. The units of the Hurwitz integers are 24 in number, and considered as points in 4-dimensional space, furnish the vertices of a polytope called the 24-cell. Unlike the Lipschitz integers, the Hurwitz integers are "well packed": If we put a ball of diameter 1 at each Hurwitz integer, they will not overlap, and there won't be room for more balls in between.

Defining an integer lattice in the octonions is rather subtle. We can of course do the Gaussian or Lipschitzian thing and take the points in $\mathbb{O}$ whose coefficients are all in $\mathbb{Z}$, but this won't be well packed. How can we do better? It is easy to slip up and define a lattice of "integers" that is not actually closed under multiplication; correcting this mistake and constructing more carefully, one finds that there are actually seven different lattices that are all isomorphic to each other! Picking one of them by convention yields the *Cayley integers,* also known as the *octavians.* The octavians have 240 elements of norm 1. Of course, this includes ± 1. Then there are 56 octavians of the form $\pm\frac{1}{2}(-1 \pm i_a \pm i_b \pm i_c)$, where the possible choices of indices a, b, c are lines in a Fano plane. There are also 126 octavians of the form $\pm i_a$ and $\frac{1}{2}(\pm i_a \pm i_b \pm i_c \pm i_d)$, where a, b, c, d are the leftovers from taking a line out of that Fano plane. (We have to tweak the labels slightly from the Fano plane that we use to define the octonion multiplication table, in order to avoid the mistake mentioned a moment ago.) Up to an overall scaling, the lattice of octavians is also the lattice known as E_8. Moreover, the octavians contain copies of all the lower-dimensional integer lattices we discussed earlier [1].

What is all this doing in a book about symmetric quantum measurements? We delve into these details because of the following surprising fact: *From the symmetries of these lattices, we can in fact read off the symmetries of the qubit, Hesse and Hoggar SICs* [2]. It is as if the sporadic SICs are drawing their strength from the octavians.[1]

To make this connection, we consider the stabilizer of the fiducial vector, i.e., the group of unitaries that map the SIC set to itself, leaving the fiducial where it is and permuting the other $d^2 - 1$ vectors. Huangjun Zhu observed that the stabilizer of any fiducial for a Hoggar-type SIC is isomorphic to the group of 3×3 unitary matrices over the finite field of order 9 [4, 5]. This group is sometimes written $U_3(3)$ or PSU(3, 3). In turn, this group is up to a factor $\mathbb{Z}_2$ isomorphic to $G_2(2)$, the automorphism group of the octavians. The stabilizer of a qubit SIC fiducial is just what we get if we mod out the unit-norm Eisenstein integers by ± 1, and in the middle, the stabilizer of a Hesse SIC fiducial is the group of unit-norm Hurwitz integers exactly.

With all this as background, we will spend this chapter investigating the Hoggar-type SICs in more detail. Section 5.4 will study the pairing of two separate eight-dimensional SICs, a pattern of interlocking geometrical relationships that will lead, in Sect. 5.5, to another application of combinatorial design theory. The results will translate to probability and information theory in Sect. 5.6, where we will see what they imply for the problem of distinguishing the consequences of different quantum-mechanical hypotheses. Investigating this further, we will arrive in Sect. 5.7 at another connection between real and complex equiangular lines.

5.2 Simplifying the QBic Equation

When in dimension $d = 8$, using a Hoggar-type SIC, the number of distinct values the triple products C_{jkl} take in this case is quite small: When the three indices are different, they can only be 0 or $\pm 1/27$.

Let S_+ denote the set of index triples (jkl) for which $C_{jkl} = 1/27$, and likewise, let S_- denote that set for which $C_{jkl} = -1/27$. We cull duplicates from these lists, so that, for example, if (jkl) belongs in S_+, we do not also include its permutations (kjl), (lkj) and so on. The sizes of these sets are

$$|S_+| = 16128 = 2^8 3^2 7, \quad |S_-| = 4032 = 2^6 3^2 7. \tag{5.3}$$

---

[1] As Baez notes, "Often you can classify some sort of gizmo, and you get a beautiful systematic list, but also some number of exceptions. Nine times out of 10 those exceptions are related to the octonions" [3]. Or perhaps we should say, seven times out of eight?.

Simplifying the QBic equation (2.20) for the special case of the Hoggar SIC proceeds by fairly straightforward algebra. The only bit of moderate cleverness required is a rearrangement by means of normalization:

$$\sum_j p(j)^2 \sum_{l \neq j} p(l) = \sum_j p(j)^2 [1 - p(j)]$$

$$= \sum_j p(j)^2 - \sum_j p(j)^3 . \tag{5.4}$$

The result of these manipulations is that a pure state must satisfy

$$\sum_j p(j)^3 + \frac{1}{3} \left[\sum_{S_+} p(j) p(k) p(l) - \sum_{S_-} p(j) p(k) p(l) \right] = \frac{11}{648} . \tag{5.5}$$

The remaining challenge is to characterize the sets S_+ and S_-.

5.3   Triple Products and Combinatorial Designs

Group covariance tells us that any C_{jkl} can be written as C_{0mn} for some m and n. This implies a d^2-fold degeneracy among the triple products. In our case, we know that the sizes of S_+ and S_- must be multiples of 64. And, in fact,

$$|S_+| = 64 \cdot 4 \cdot 3^2 7, \quad |S_-| = 64 \cdot 3^2 7. \tag{5.6}$$

In forming the triple product C_{0mn}, we have

$$\binom{63}{2} = \frac{63 \cdot 62}{2} = 1953 \tag{5.7}$$

ways of choosing the subscripts m and n. We find that

$$\frac{|S_-|}{64} = \frac{1}{31} \binom{63}{2}, \quad \frac{|S_+|}{64} = \frac{4}{31} \binom{63}{2}. \tag{5.8}$$

It is now time to go into the group theory of SIC structures in more detail. We define the *multipartite Weyl–Heisenberg group* in a prime-power dimension p^n to be the tensor product of n copies of the Weyl–Heisenberg group in dimension p. The corresponding Clifford group in dimension p^n is the group of unitaries that stabilize the multipartite Weyl–Heisenberg group. The order of this Clifford group [5] is

$$p^{n^2+2n} \prod_{j=1}^{n} (p^{2j} - 1) .$$ (5.9)

Therefore, in dimension $8 = 2^3$, the Clifford group has order

$$2^{3^2+2\cdot3} \prod_{j=1}^{3} (2^{2j} - 1) = 2^{15} \cdot 3 \cdot 15 \cdot 63 = 2^{15} \cdot 3^4 \cdot 5 \cdot 7$$

$$= 2^9 \cdot 3^2 \cdot (|S_+| + |S_-|) .$$ (5.10)

The symmetry group of a Hoggar SIC is a subgroup of the Clifford group with order

$$64 \cdot 6048 = 2^{11} \cdot 3^3 \cdot 7 = 24|S_+| = 96|S_-| .$$ (5.11)

The factor of $64 = 2^6$ comes from the triple-qubit Pauli group. Take any vector from the Hoggar SIC, and consider those unitaries in the symmetry group that leave that vector fixed while permuting the others. These form the stabilizer subgroup of that vector. For any vector in the Hoggar SIC, the stabilizer subgroup is isomorphic to the projective special unitary group PSU(3, 3), which has 6048 elements. This explains the other factor in Eq. (5.11).[2]

Let N_k^+ be the number of triples in the set S_+ that contain the value k, and likewise for N_k^- and S_-. One finds that

$$N_k^- = 189, \ N_k^+ = 756 \ \forall k .$$ (5.12)

These values factorize as

$$N_k^- = 3^3 \cdot 7, \ N_k^+ = 2^2 \cdot 3^3 \cdot 7 .$$ (5.13)

Furthermore, if we let $N_{kl}^{\pm}$ denote the number of triples in S_+ (respectively, S_-) that contain the pair (k, l), we obtain

$$N_{kl}^- = 6, \ N_{kl}^+ = 24, \ \forall k, l .$$ (5.14)

This leads us into *combinatorial design theory*. A *balanced incomplete block design* (BIBD) is a collection of v points and b blocks, such that there are k points within each block, and r blocks contain any given point. Consistency requires that

$$bk = vr .$$ (5.15)

---

[2]If one constructs the Hoggar SIC as Zhu does, then its fiducial vector's stabilizer group is generated by his unitary operators U_7 and U_{12}. Construct the new unitaries $U_a = U_{12}U_7$ and $U_b = U_{12}^2$. These satisfy the relations for the generators of PSU(3, 3) as presented in the [6]. Also, the conjugacy classes in Zhu's Table 10.1 can be matched with those for PSU(3, 3) computed, for example, using the GAP software [7].

The final parameter, λ, specifies the number of blocks containing any two specific points. This constant must satisfy

$$\lambda(v - 1) = r(k - 1). \tag{5.16}$$

In a *symmetric design*, $b = v$, and so $r = k$. Any two blocks meet in the same number of points, and that number is λ. Ryser's theorem [8] establishes that this is an if-and-only-if relationship.

The set S_- contains 4032 "blocks," where each block is made of three points drawn from a set of 64 possibilities. We found earlier that each point occurs in 189 different blocks, and that each pair of points occurs in 6 different blocks. Therefore, S_- is a BIBD with

$$v = 64, \ \ b_- = |S_-| = 4032, \ \ k = 3, \ \ r_- = 189, \ \ \lambda_- = 6. \tag{5.17}$$

Likewise, S_+ is a BIBD with

$$v = 64, \ \ b_+ = |S_+| = 16128, \ \ k = 3, \ \ r_+ = 756, \ \ \lambda_+ = 24. \tag{5.18}$$

Referring back to Eq. (5.11), we have that

$$b_\pm = |S_\pm| = \frac{6048v}{4\lambda_\mp} = \frac{6048v}{4} \frac{\lambda_\pm}{\lambda_-\lambda_+} = \frac{6048v}{576}\lambda_\pm = 672\lambda_\pm. \tag{5.19}$$

As mentioned above, Zhu proved that the Hoggar SIC is doubly transitive, i.e., for any distinct pair of vectors, there is a symmetry operation that takes it to any other distinct pair [5]. This has implications for the structure coefficient matrices C_i, defined by

$$(C_i)_{jk} = C_{ijk}. \tag{5.20}$$

Group covariance means that

$$C_{ijk} = C_{0j'k'} \tag{5.21}$$

for some j' and k'. So, the entries in all the matrices $\{C_i\}$ are elements of the matrix C_0. The additional requirement that the action of the symmetry group is doubly transitive means that if we want to understand the triple products C_{ijk}, we only need to look at C_{01k}, because any triple of distinct indices (ijk) can be mapped to some $(01k')$, leaving the triple product invariant.

We expect to see some values occur in sets of six, or multiples of six. Why? Because the triple product function is completely symmetric:

$$C_{ijk} = C_{jki} = C_{kij} = C_{jik} = C_{kji} = C_{ikj}. \tag{5.22}$$

By applying unitaries in the symmetry group, we can turn the first pair of indices into ij across the board:

$$C_{ijk} = C_{ij\sigma_1(i)} = C_{ij\sigma_2(j)} = C_{ij\sigma_3(k)} = C_{ij\sigma_4(i)} = C_{ij\sigma_5(j)} . \tag{5.23}$$

Here, the $\{\sigma_1, \ldots, \sigma_5\}$ are permutations of the set of indices $\{0, \ldots, 63\}$. They are defined by relations of the form

$$\sigma_1(j) = i, \ \sigma_1(k) = j . \tag{5.24}$$

Unless these permutations happen to align in such a way that, for example, $\sigma_4(i) = \sigma_5(j)$, we will have six elements in the jth row of the matrix C_i, all equal.

Explicit computation bears this idea out. We need the values of C_{01k}, where the subscripts "0" and "1" refer to the first and second projectors in the ordering defined by Eq. (2.38). Note that two entries will be the trivial value, $1/(d+1)$. We can display the results by arranging them in a $4 \times 4 \times 4$ cube. Define the sequence

$$\sigma = \{I, \sigma_z, \sigma_x, \sigma_x\sigma_z\} . \tag{5.25}$$

Then, interpreting the index k as an ordered tuple $(k_0, k_1, \ldots, k_5)$, we have

$$C_{01k} = \operatorname{Re} \operatorname{tr}(\Pi_0 \Pi_1 D_k \Pi_0 D_k^\dagger) , \tag{5.26}$$

where

$$D_k = \sigma_{k_1+2k_0} \otimes \sigma_{k_3+2k_2} \otimes \sigma_{k_5+2k_4} . \tag{5.27}$$

We can therefore display C_{01k} for all k in a three-dimensional cube, which is portrayed in Fig. 5.1.

The pairing of values follows from the facts that $C_{01k} = C_{10k}$ by symmetry and

$$D_1^2 = (I \otimes I \otimes Z)^2 = I \otimes I \otimes I . \tag{5.28}$$

This makes the triple product insensitive to a Z factor on one qubit. However, if the displacement operator includes a factor of X on that qubit, then the triple product C_{01k} vanishes. Inspection reveals that among the nonvanishing values, $C_{01k} = -1/27$ when the displacement operator D_k includes only factors of X, apart from the third qubit, which is insensitive to Z.

Define the complex triple products

$$T_{jkl} = \langle \psi_j | \psi_k \rangle \langle \psi_k | \psi_l \rangle \langle \psi_l | \psi_j \rangle = \operatorname{tr}(\Pi_j \Pi_k \Pi_l) . \tag{5.29}$$

Up until now, we have taken the real part of this quantity. We can instead scale by the magnitude to obtain a phase [9]:

$$\tilde{T}_{jkl} = \frac{T_{jkl}}{|T_{jkl}|} = e^{i\theta_{jkl}} . \tag{5.30}$$

It follows from the definition of T_{jkl} that, in general,

Fig. 5.1 Visual representation of C_{01k} for the Hoggar SIC. Small dots indicate $C_{01k} = 0$. Large spheres (red) indicate the trivial value, $C_{01k} = 1/9$. Intermediate spheres (yellow) indicate $C_{01k} = 1/27$, and the six slightly smaller spheres (blue) stand for $C_{01k} = -1/27$

$$e^{i\theta_{mjk}} e^{i\theta_{mkl}} e^{i\theta_{mlj}} = e^{i\theta_{jkl}} . \tag{5.31}$$

For the Hoggar SIC, θ_{jkl} takes the values 0, π and $\pm\pi/2$.

Note that the definition of a SIC implies that

$$\langle \psi_j | \psi_k \rangle = \frac{1}{\sqrt{d+1}} e^{i\theta_{jk}} \tag{5.32}$$

for some angles θ_{jk}. This two-index object is related to the three-index object θ_{jkl} by

$$e^{i\theta_{jkl}} = e^{i\theta_{jk}} e^{i\theta_{kl}} e^{i\theta_{lj}} . \tag{5.33}$$

With this relation, we can understand more about the triple products C_{jkl} using the following sneaky trick. The operators X and Z are Hermitian, but XZ is not. We can fix this by defining

$$Y = iXZ , \tag{5.34}$$

which is a Hermitian operator (and equal to the familiar Pauli matrix σ_y). A tensor-product operator like $X \otimes Z \otimes XZ$ will not be Hermitian, but $X \otimes Z \otimes Y$ will be. So, by introducing appropriate phase factors, we can fix up the Weyl–Heisenberg displacement operators D_k so that they are Hermitian matrices. The phase with which we modify D_k includes a factor of i for every instance of Y in the tensor product:

$$\hat{D}_k = (-e^{i\pi/d})^{\#(Y)} D_k . \tag{5.35}$$

These operators serve just as well for generating a SIC.

But notice: Our displacement operators are now Hermitian matrices, that is, *quantum observables,* and their expectation values are real. Consequently, for any $\hat{D}_k$,

$$\langle \psi_0 | \hat{D}_k | \psi_0 \rangle \in \mathbb{R} . \tag{5.36}$$

In turn, this implies that

$$e^{i\theta_{0k}} = \pm 1 . \tag{5.37}$$

Denote by S_0 the set of all triples (jkl) for which C_{jkl} vanishes. For these triples, it must be the case that T_{jkl} is pure imaginary. Let us focus on the case $j = 0$, with $k \neq 0$ and $l \neq 0$. Here, the only place a factor of i can enter is the middle:

$$e^{i\theta_{0kl}} = e^{i\theta_{0k}} e^{i\theta_{kl}} e^{i\theta_{l0}} . \tag{5.38}$$

The middle factor is the phase of the inner product

$$\langle \psi_k | \psi_l \rangle = \langle \psi_0 | \hat{D}_k^\dagger \hat{D}_l | \psi_0 \rangle . \tag{5.39}$$

This can yield an imaginary part for some values of k and l, thanks to the phase factors we introduced to obtain Hermiticity. Write $\{\cdot, \cdot\}$ for the symplectic form

$$\{a, b\} = a_1 b_0 - b_1 a_0 . \tag{5.40}$$

Then the phase we obtain is

$$(-i)^{\{(k_0,k_1),(l_0,l_1)\}+\{(k_2,k_3),(l_2,l_3)\}+\{(k_4,k_5),(l_4,l_5)\}} . \tag{5.41}$$

If we fix the index k, say to

$$(k_0, k_1, k_2, k_3, k_4, k_5) = (0, 0, 0, 0, 0, 1) , \tag{5.42}$$

then the phase contribution will be an imaginary number for exactly 32 of the 64 possible choices of the index l. These are the values for which $C_{01l} = 0$. If we define the matrix

$$\Omega = I_{3\times 3} \otimes \begin{pmatrix} 0 & -1 \\ 1 & 0 \end{pmatrix} , \tag{5.43}$$

then we can write the exponent in Eq. (5.41) as

$$f(k, l) = k \Omega l^{\mathsf{T}} , \tag{5.44}$$

where we are interpreting k and l as row vectors of six elements each. The matrix Ω is invertible and antisymmetric, so $f(k, l)$ is a *symplectic bilinear form.*

Let us consider again the three-index angle tensor θ_{jkl}. We know that

$$e^{i\theta_{jkl}} = \pm i, \text{ for } (jkl) \in S_0. \tag{5.45}$$

If (mjk), (mkl) and (mlj) are three triples in S_0, then

$$e^{i\theta_{jkl}} = \pm i. \tag{5.46}$$

That is, (jkl) must then be a member of S_0, too. On the other hand, if (mjk), (mkl) and (mlj) are all *outside* of S_0, then $e^{i\theta_{jkl}}$ is the product of three real numbers, and so it must be real itself. Therefore, if (mjk), (mkl) and (mlj) are in the complement of S_0, then so is (jkl).

This means that S_0 qualifies as a *two-graph.* Much studied in discrete mathematics, a two-graph can be defined [10] as a set T of triples such that

$$(pqr), (pqs), (prs) \in T \Rightarrow (qrs) \in T, \tag{5.47}$$

and likewise for the complement of T.

One application of two-graphs is generating sets of equiangular lines in *real* vector spaces. Pick a point in $\mathbb{R}^d$, and draw a set of lines through it, such that any two meet at an angle whose cosine is $\pm\alpha$ (with $\alpha \neq 0$). For some triples of those intersecting lines, the product of the cosines will be negative, and for others, it will be positive. The triples for which the product is negative constitute a two-graph. Going in the other direction, any two-graph can be formulated in this way.

Notice what has happened here: We started with a set of *complex* equiangular lines, the Hoggar SIC, and in considering the additional symmetries that set enjoys above and beyond its definition, we have arrived at *real* equiangular lines.

This will happen again.

Two-graphs have been taxonomied to an extent, with the aid of the classification theorem for finite simple groups. Those two-graphs with doubly transitive automorphism groups were classified by Taylor [11]. Our set S_0 is Taylor's example B.xi, the two-graph on 64 vertices whose automorphism group contains PSU(3, 3).

Knowing the automorphism group of this two-graph, we can find the stabilizer of any pair of vertices. This will be the subgroup whose action leaves that pair fixed. For example, automorphisms in the stabilizer subgroup of the pair (0, 1) will send the triple $(01k)$ to the triple $(01k')$. Taylor [11] observes that the stabilizer of two points in a triple has orbits of length 6, 24 and 32 on the remaining points. Combining this with Zhu's observation [4] that two triples in the Hoggar SIC can be mapped to each other by a symmetry operation if and only if they have the same triple product, and we see a combinatorial origin of the patterns we observed in Fig. 5.1.

Given a two-graph T, one can construct a regular graph G that embodies its structure in the following manner [11]. Copy over the list of vertices from T to G. Then, select a vertex v of the two-graph T, and draw the edges of G so that u and w are neighbors whenever $(uvw) \in T$. Let A be the *Seidel adjacency matrix* of the graph G. This matrix is constructed so that $A_{uw} = -1$ if u and w are adjacent, $A_{uw} = 1$ if they are not, and $A_{uu} = 0$ on the diagonal. Suppose that the smallest eigenvalue of A is λ, and this eigenvalue occurs with multiplicity m. Then, $M = I - (1/\lambda)A$ is a symmetric, positive definite matrix, and the rank of M will be the number $|A|$ of vertices in the graph minus the multiplicity m. Consequently, M can be taken as the Gram matrix for a set of vectors

$$\{v_1, v_2, \ldots, v_{|A|}\}, \tag{5.48}$$

with each vector living in $\mathbb{R}^{|A|-m}$.

In our case, the matrix A has only two eigenvalues: 7, with multiplicity 36; and -9, with multiplicity 28. This means that M is the Gram matrix for a set of *equiangular lines* (as it should be, since we derived G from a two-graph).

From the triple-product structure of the Hoggar SIC, we have arrived at a set of 64 equiangular lines in $\mathbb{R}^{36}$.

The numbers 28 and 36 will recur in the next developments.

5.4 The Twin of the Hoggar SIC

Now, we investigate the eight-dimensional analogue of what happens when we minimize the Shannon entropy for qubit pure states.

The "twin Hoggar SIC" can be constructed by applying the triple-Pauli displacement operators to the fiducial vector

$$\left| \tilde{\psi}_0 \right\rangle \propto (-1 - 2i, 1, 1, 1, 1, 1, 1, 1)^{\mathsf{T}}. \tag{5.49}$$

This is related to our original fiducial vector, Eq. (2.37), by complex conjugation.

In the SIC representation defined by the original Hoggar lines, the vectors comprising the "twin Hoggar SIC" have $(8 - 1)8/2 = 28$ elements equal to zero, and the other $(8 + 1)8/2 = 36$ elements equal to $1/36$ [12]. Consequently, the Hoggar lines provide a counterexample to the conjecture that the best upper bound on the number of zero-valued entries in dimension d is just d. The bound $d(d - 1)/2$ deduced from the Cauchy–Schwarz inequality [13] is, actually, tight. Furthermore, the states of the twin Hoggar SIC minimize the Shannon entropy of their SIC representations, as we discussed above. One can, in fact, find the the twin Hoggar SIC-set by testing all the states of the form (2.27) to see which ones satisfy the QBic equation.

For any vector p in the twin Hoggar SIC set,

$$\sum_j p(j)^3 = 36 \left(\frac{1}{36}\right)^2 = \frac{1}{1296}. \tag{5.50}$$

Equation (5.5) then becomes

$$\frac{1}{1296} + \frac{1}{3}\left[\sum_{S_+} p(j)p(k)p(l) - \sum_{S_-} p(j)p(k)p(l)\right] = \frac{11}{648}. \tag{5.51}$$

The bracketed sum must therefore equal

$$\left[\sum_{S_+} p(j)p(k)p(l) - \sum_{S_-} p(j)p(k)p(l)\right] = \frac{7}{144}. \tag{5.52}$$

Furthermore, any product $p(j)p(k)p(l)$ that does not evaluate to zero must equal

$$p(j)p(k)p(l) = \left(\frac{1}{36}\right)^3 = \frac{1}{46,656}. \tag{5.53}$$

From this, we can calculate the net number of contributions that the sums over S_+ and S_- must make, if the state is to be valid:

$$\frac{\frac{7}{144}}{\frac{1}{46,656}} = 2,268 = 2^2 3^4 7 = \frac{3(|S_+| - |S_-|)}{2^4}. \tag{5.54}$$

If π_i and π_j are two projectors in the twin set, then

$$\mathrm{tr}(\pi_i \pi_j) = d(d+1)\sum_k p_i(k)p_j(k) - 1 = \frac{1}{d+1}. \tag{5.55}$$

Therefore,

$$\sum_k p_i(k)p_j(k) = \frac{d+2}{d(d+1)^2} = \frac{5}{324}. \tag{5.56}$$

Now, each element in p_i is either 0 or $1/36$, and likewise for p_j. Let n denote the number of overlapping nonzero entries in these two vectors. We know that

$$n\left(\frac{1}{36}\right)^2 = \frac{5}{324}, \tag{5.57}$$

and so

$$n = 20. \tag{5.58}$$

This result will be important for understanding the twin Hoggar SIC using combinatorial design theory.

Because we have 28 vectors in $\mathbb{C}^8$ that are all orthogonal to a common vector, they fit into $\mathbb{C}^7$. Iverson and Mixon have proved that for any q, there exists a set of $q^3 + 1$ complex equiangular lines living in $d = q^2 - q + 1$ dimensions whose automorphism group is doubly transitive and is isomorphic to $\mathrm{PSU}(3, q)$ or has it as a subgroup [14]. Our 28 lines are the $q = 3$ case.

5.5   Combinatorial Designs from the Twin Hoggar SIC

We have a set of $d^2 = 64$ "blocks", each one of which essentially is a binary string of length 64. And each block contains exactly 36 of the nonzero entries that a length-64 block could in principle contain. We can think of this as there being 64 "points," and each block contains 36 of them. Table 6.1 gives examples of four such blocks.

If we fix $v = b = 64$ and $k = 36$, then

$$\lambda \cdot 63 = 36 \cdot 35 \;\; \Rightarrow \;\; \lambda = 20 \,. \tag{5.59}$$

This is just what we found before when we calculated the number of overlapping 1s in any pair of vectors in the twin Hoggar set. Therefore, the twin Hoggar SIC defines a symmetric design. Specifically, it is a "2-(64,36,20) design".

If we apply a NOT to each of our bit-strings, then we arrive at a new design. Generally, the *complement* of a design is found by replacing each block with its complement: The points that were included in a block are now excluded, and vice versa. The new design has parameters

$$v' = v, \; b' = b, \; k' = v - k, \; r' = b - r, \; \lambda' = \lambda + b - 2r \,. \tag{5.60}$$

The complement to our Hoggar design therefore satisfies

$$v' = b' = 64, \; k' = r' = 28, \; \lambda' = 12 \,. \tag{5.61}$$

Therefore, we can designate it a "2-(64,28,12) design."

The existence of a symmetric design with parameters

$$(v, k, \lambda) = (4u^2, 2u^2 - u, u^2 - u) \tag{5.62}$$

is known to be equivalent to the existence of a regular Hadamard matrix possessing dimensions $4u \times 4u$. Setting $u = 4$, we find that the complement of the Hoggar design meets the Hadamard criterion. The incidence matrix of the design can be transformed into a regular Hadamard matrix by simple substitutions.

The complement of the Hoggar design is equivalent to an *orthogonality graph* for the Hoggar SIC and its twin. In an orthogonality graph, vertices stand for states, and vertices are linked by an edge if the corresponding states are orthogonal. If a point V_i lies within block B_j, then the ith vector in the Hoggar SIC is orthogonal to the

jth vector in the twin SIC. This can be visualized as a bipartite graph containing two sets of 64 vertices apiece, where each vertex in the first set is linked to 28 vertices in the second set.

We can generate the Hoggar design in another way by the following procedure. Start with this Hadamard matrix:

$$H_2 = \begin{pmatrix} -1 & 1 & 1 & 1 \\ 1 & -1 & 1 & 1 \\ 1 & 1 & -1 & 1 \\ 1 & 1 & 1 & -1 \end{pmatrix}. \tag{5.63}$$

Construct the tensor product of three copies of H_2:

$$H_6 = H_2 \otimes H_2 \otimes H_2. \tag{5.64}$$

Then, use this to create an incidence matrix by replacing all the entries that equal -1 with 0:

$$M = \frac{H_6 + 1}{2}. \tag{5.65}$$

The resulting 64×64 array is the incidence matrix of the Hoggar design, containing all the same rows as the (appropriately renormalized) SIC representations of the twin set. This ties us firmly into the literature on combinatorial designs: The Hoggar design is a *symplectic design on 64 points*.[3] A symplectic design [15–17], denoted $S^\epsilon(2m)$ with m a positive integer and $\epsilon = \pm 1$, is a BIBD with

$$b = v = 2^{2m}, \quad k = 2^{2m-1} + \epsilon 2^{m-1}, \quad \lambda = 2^{2m-2} + \epsilon 2^{m-1}. \tag{5.66}$$

The object that we found by way of SIC-POVMs is exactly $S^1(2m)$ for $m = 3$. Symplectic designs for larger m can be constructed by taking the tensor product of m copies of the Hadamard matrix H_2.

That is how to construct the symplectic designs $S^\pm(6)$, as combinatorial geometries. Does the matrix H_2 have a meaning in quantum physics? In fact, it does. In qubit state space, a SIC is a regular tetrahedron inscribed within the Bloch sphere. Finding the minimum-entropy pure states, as we did for the Hoggar SIC, they turn out to form a second tetrahedron, dual to the first. Together, the two SICs constitute a stellated octahedron in the Bloch-sphere representation. Each projector in the new SIC is orthogonal to exactly one of the four projectors in the original SIC. Let $J_{4 \times 4}$ be the 4×4 matrix whose entries are all 1. Then, up to normalization, the SIC representations of the four new projectors can be written as the rows of the matrix

---

[3] While these notes were in preparation, Szymusiak and Słomczyński updated an earlier arXiv paper of theirs with an independent derivation of this point [12].

$$
\begin{pmatrix} 0\ 1\ 1\ 1 \\ 1\ 0\ 1\ 1 \\ 1\ 1\ 0\ 1 \\ 1\ 1\ 1\ 0 \end{pmatrix} = J_{4\times 4} - I_{4\times 4}. \tag{5.67}
$$

This is clearly just the Hadamard matrix H_2, shifted and rescaled. So, the structure of orthogonalities between the Hoggar SIC and its twin is, essentially, the tensor product of three copies of the analogous structure for a qubit SIC.

An automorphism of a symmetric design is a permutation of the points that preserves the block structure, sending blocks to blocks. The symplectic designs admit *2-transitive automorphism groups*. That is, the automorphism group of a symplectic design $S^\epsilon(2m)$ contains permutations that map any pair of points to any other pair of points. Furthermore, the automorphism group of a design is 2-transitive for points if and only if it is so for blocks as well. Therefore, the automorphism group of a symplectic design includes transformations that can map any pair of blocks to any other pair of blocks.

The *symmetric difference* of two sets is defined to be the set of those elements contained in their union but not their intersection. For example,

$$
\{\text{Alice, Bob, Charlie}\} \ominus \{\text{Charlie, Demona}\}
$$
$$
= \{\text{Alice, Bob, Demona}\}. \tag{5.68}
$$

If the symmetric difference of any *three* blocks in a design is either a block or the complement of a block, then that design is said to have the *symmetric difference property*. If a design enjoys the symmetric difference property, then that design or its complement meets the following condition [18] on its parameters:

$$
v = 2^{2m}, \quad k = 2^{2m-1} - 2^{m-1}, \quad \lambda = 2^{2m-2} - 2^{m-1}. \tag{5.69}
$$

The complement of the Hoggar design satisfies these conditions with $m = 3$.

The 64-point designs with the symmetric difference property can be completely classified [18]. There exist four inequivalent such designs, distinguished by their automorphism groups [16]. The symplectic design, which we found by way of the Hoggar SIC, is the most symmetric: It is the only one of the four whose automorphism group is 2-transitive.

Let $\mathbb{F}_2$ denote the finite field of order two, and let $\mathrm{Sp}(2m, \mathbb{F})$ denote the group of $2m \times 2m$ symplectic matrices over the field $\mathbb{F}$. Then, the automorphism group of the symplectic design $S^1(6)$ is isomorphic to

$$
G = (\mathbb{Z}_2)^6 \times \mathrm{Sp}(6, \mathbb{F}_2). \tag{5.70}
$$

The stabilizer of any point is $\mathrm{Sp}(6, \mathbb{F}_2)$.

The original SIC and the twin SIC have the same symmetry group. Let Π_i be a projector in the original set and π_i a projector in the twin set. Suppose that g is an element of the symmetry group that takes π_j to π_k. Then

$$\text{tr}(\Pi_i \pi_k) = \text{tr}(\Pi_i g \pi_j g^\dagger) = \text{tr}(g^\dagger \Pi_i g \pi_j). \tag{5.71}$$

So, the SIC representation of π_j is just the SIC representation of π_k, with the entries permuted. Any element of the Hoggar SIC's symmetry group corresponds to a permutation that preserves the combinatorial design structure. However, the converse is not true: Not all elements in the automorphism group G can be implemented by unitaries that belong to the Hoggar SIC's symmetry group. This is a restatement of the fact that the symmetry group of the Hoggar SIC is a proper subgroup of the triple-qubit Clifford group.

5.6 Quantum-State Compatibility

A quantum state can be thought of as a hypothesis for how a quantum system will behave when experimented upon. When are two such hypotheses different in a meaningful way? One way of quantifying this is the idea of *compatibility* between quantum states. Two quantum states ρ and ρ' are what we will call *P-incompatible* if a measurement exists that meets the following condition [19]. Let the measurement outcomes be labeled by j, so that the operators $\{E_j\}$ form a POVM,

$$\sum_j E_j = I. \tag{5.72}$$

The probabilities for the outcomes are computed using the Born rule:

$$q(j) = \text{tr}(\rho E_j), \ q'(j) = \text{tr}(\rho' E_j). \tag{5.73}$$

If one can devise a measurement $\{E_j\}$ such that for *any* outcome j, at least one of $q(j)$ or $q'(j)$ is zero, then the states ρ and ρ' are P-incompatible. Otherwise, the states are P-compatible. This can naturally be generalized to the question of compatibility among three or more states. Special cases of P-incompatibility are *O-incompatibility,* where the measurement is an orthonormal basis (à la von Neumann), and *S-incompatibility,* where it is a SIC.[4]

Is it possible for quantum states to be P-incompatible with respect to a SIC measurement? Yes, but not if we only consider two states at a time. For example, these are three valid states for the Hesse SIC representation in dimension $d = 3$.

---

[4] Our terminology for types of compatibility is nonstandard. The paper which originally codified these criteria [19] used jawbreaker acronyms, and later writings in the area invented jargon that is not much better.

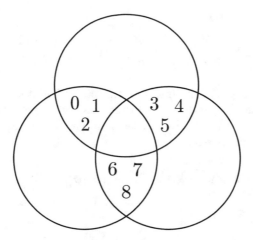

Fig. 5.2 Pictorial representation of the hypotheses defined in Eq. (5.74). Each circle corresponds to a quantum state. The numbers indicate the outcomes that are consistent with that state, i.e., the outcomes for which that state implies nonzero probability

$$\left(0, 0, 0, \frac{1}{6}, \frac{1}{6}, \frac{1}{6}, \frac{1}{6}, \frac{1}{6}, \frac{1}{6}\right);$$

$$\left(\frac{1}{6}, \frac{1}{6}, \frac{1}{6}, 0, 0, 0, \frac{1}{6}, \frac{1}{6}, \frac{1}{6}\right);$$ (5.74)

$$\left(\frac{1}{6}, \frac{1}{6}, \frac{1}{6}, \frac{1}{6}, \frac{1}{6}, \frac{1}{6}, 0, 0, 0\right).$$

Note that there is exactly one zero in each column. In other words, for each outcome of the Hesse SIC, exactly one of these three states assigns that outcome a probability of zero.

This is a situation where the relationship among three entities is not clearly apparent from the relationships within each pair. In such a case, it can be helpful to portray the configuration diagramatically [20–22]. We do so in Fig. 5.2. Each circle in Fig. 5.2 stands for one of the three states given in Eq. (5.74). The numbers contained within a circle are the labels of the outcomes that are consistent with that state. Note that these outcomes are only written in the areas where two circles overlap. No outcome belongs to a single state alone, and no outcome belongs to all three.

Suppose we have three pure states in dimension $d = 3$. We denote them by $|\psi\rangle$, $|\psi'\rangle$ and $|\psi''\rangle$. These can be considered as three different hypotheses that an agent Alice is willing to entertain about a quantum system. If they are P-incompatible, then there exists some measurement that Alice can perform such that for any outcome of that measurement, at least one of the three hypotheses deems that outcome impossible. A necessary and sufficient condition [23, 24] for three pure states in $d = 3$ to be O-incompatible is for the following inequalities to be satisfied. First,

$$|\langle\psi|\psi'\rangle|^2 + |\langle\psi'|\psi''\rangle|^2 + |\langle\psi''|\psi\rangle|^2 < 1,$$ (5.75)

and second,

$$\left(|\langle \psi | \psi' \rangle|^2 + |\langle \psi' | \psi'' \rangle|^2 + |\langle \psi'' | \psi \rangle|^2 - 1 \right)^2 \tag{5.76}$$
$$\geq 4 |\langle \psi | \psi' \rangle|^2 |\langle \psi' | \psi'' \rangle|^2 |\langle \psi'' | \psi \rangle|^2 .$$

Consider what happens if the three states are drawn from a SIC set. No set of three vectors can span more than three dimensions, so even though our states naturally live in a higher-dimensional Hilbert space, we press forward and use the three-dimensional criterion. In that case,

$$|\langle \psi | \psi' \rangle|^2 = |\langle \psi' | \psi'' \rangle|^2 = |\langle \psi'' | \psi \rangle|^2 = \frac{1}{d+1} . \tag{5.77}$$

The first inequality becomes

$$\frac{3}{d+1} < 1 , \tag{5.78}$$

and the second inequality becomes

$$\left(\frac{3}{d+1} - 1 \right)^2 \geq \frac{4}{(d+1)^3} . \tag{5.79}$$

We can simplify the latter expression to

$$(d-2)^2 \geq \frac{4}{d+1} . \tag{5.80}$$

Both inequalities are satisfied simultaneously for $d \geq 3$.

The three-dimensional criterion tells us that there is *some* von Neumann measurement with respect to which the three states drawn from the SIC set are incompatible. Yet the SIC is itself a measurement, and with respect to *that* measurement, the three states are entirely compatible. In the representation that the SIC itself defines, the states $|\psi\rangle$, $|\psi'\rangle$ and $|\psi''\rangle$ all have the form

$$e_k(i) = \frac{1}{d(d+1)} + \frac{1}{d+1} \delta_{ik} \tag{5.81}$$

for some values of k, and these vectors contain no zeros at all. We have here a rather cute situation. Classically, an "informationally complete measurement" would be something like an experiment that discovers the exact values of all the positions and momenta of the particles comprising a system. We tend to think of any other measurement as a coarse-graining of that one, a measurement that throws away some of the information that is, in principle, available. And throwing away information makes classical configurations *harder* to distinguish from one another. If we can rule out a hypothesis using a clumsy, imprecise measurement, then *surely* we could do so using a *maximally informative* one! How could a measurement that is less exhaustive be *better* at ruling out a hypothesis?

This is indicative of the way in which quantum physics runs counter to classical intuition. An informationally complete quantum measurement is *not* the determination of the values of all hidden variables, or the narrowing of a Liouville density to a delta function. A vector in a SIC representation is not a probability distribution over a putative hidden-variable configuration space. And we do not calculate the probabilities for outcomes of other experiments merely by blurring over IC ones.

The double-transitivity of the Hoggar SIC simplifies the structure of the triple products, as we saw above. It does the same for considerations of P-compatibility [23], as well.

Any two SIC vectors are P-compatible. However, a set of three SIC vectors when taken together can be P-incompatible. In dimension 3, the measurements that reveal P-incompatibility for the Hesse SIC are a collection of vectors originally known for other reasons: They comprise four Mutually Unbiased Bases (MUB) [24]. What about with the Hoggar SIC?

Use one set of Hoggar lines to define a SIC representation of state space, and translate the twin Hoggar lines into this representation. Any two projectors in the twin Hoggar set will be pairwise P-compatible. Direct computation shows that any set of three distinct projectors will also be compatible, in the sense that the Hoggar SIC measurement itself will not reveal any incompatibility. However, a set of four lines from the twin Hoggar set can be S-incompatible, with that incompatibility revealed by the original Hoggar SIC-POVM itself. For example, in Table 6.1 we gave the SIC representations of four lines from the twin Hoggar set.

As we noted earlier, a "1" in a bitstring means that the entry in that place is the nonzero value appropriate for the dimension, which here is $1/36$. A "0" means that the vector is zero in that slot.

There is at least one 0 in each column, meaning that for every possible outcome of the Hoggar SIC-POVM, one of these four state assignments deems that outcome impossible. However, if we leave out any of the four rows, this is no longer true.

We illustrate this in Fig. 5.3. Each ellipse stands for a state vector, that is, for a row in Table 6.1. The central region, where all four ellipses overlap, contains no outcomes.

There are lots of other examples; we do, after all, have 64 vectors to choose from. However, we should be able to simply the problem, and understand what's going on by considering only a subset of all those combinations. Why?

If we apply the same permutation to the four rows shown above, the columns still line up, meaning that there is still at least one zero in each column. Consequently, the transformed states will also be S-incompatible.

Because we can take any distinct pair (π_j, π_k) to the pair (π_0, π_1), then we should be able to understand the S-compatibility properties of all quadruples by working out what happens with $(\pi_0, \pi_1, \pi_m, \pi_n)$.

We now apply our knowledge of combinatorial design theory. Let B_i denote the bitstring representation of the state π_i in the twin Hoggar SIC. These 64 sequences, which we can think of as the rows in a square matrix, form a symmetric design, as we showed earlier, and this design has the symmetric difference property. In terms of bitstrings, the symmetric difference $B_i \ominus B_j$ is equivalent to an XOR operation:

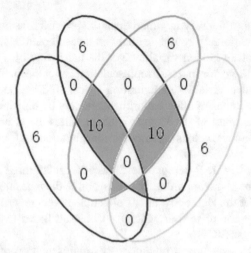

Fig. 5.3 Venn diagram for the set of four states from the twin Hoggar SIC given in Table 6.1. Each ellipse stands for a quantum state. Labels indicate the number of outcomes of the Hoggar SIC for which that state implies nonzero probability. The shaded regions, where exactly three of the four ellipses overlap, contain 10 outcomes. Each ellipse contains three such regions, as well as a region all to itself. In total, each ellipse contains a value of 36. The central region, where all four ellipses intersect, contains 0. (Figure based on [25].)

Table 5.1 The XOR of three bits, and its complement

a	b	c	a XOR b XOR c	NOT$(a$ XOR b XOR $c)$
0	0	0	0	1
0	0	1	1	0
0	1	0	1	0
0	1	1	0	1
1	0	0	1	0
1	0	1	0	1
1	1	0	0	1
1	1	1	1	0

$$(B_i \ominus B_j)(n) = B_i(n) \text{ XOR } B_j(n) . \tag{5.82}$$

This is readily verified, and implies the convenient fact that the symmetric difference is associative:

$$(B_i \ominus B_j) \ominus B_k = B_i \ominus (B_j \ominus B_k) . \tag{5.83}$$

In Table 5.1, we show the values resulting from applying XOR symmetrically to three bits, and the complementary values.

Because the Hoggar design has the symmetric difference property, the symmetric difference of any three blocks is either a block or the complement of a block. Suppose that the symmetric difference of B_i, B_j and B_k is the complement of B_l. Then B_l is the complement of $B_i \ominus B_j \ominus B_k$. We can find each element of B_l by locating the proper row in Table 5.1. It follows that for all $n \in \{0, \ldots, 63\}$, the set

$$\{B_i(n), B_j(n), B_k(n), B_l(n)\} \tag{5.84}$$

contains either 1, 2 or 3 zeroes. That is, these elements are never all zero, nor are they all ever one. Consequently, a measurement of the Hoggar SIC-POVM reveals S-incompatibility among the four states $\{\pi_i, \pi_j, \pi_k, \pi_l\}$ in the twin Hoggar SIC.

What else can we say about the symmetric differences of the blocks $\{B_i\}$? Each $B_i \ominus B_j$ for a distinct pair $i \neq j$ is a list of positions where exactly one of $B_i(n)$ and $B_j(n)$ equals one. By direct computation, we find that each such list is 32 items long. We can pick a pair of distinct blocks in 2,016 different ways. However, not all choices yield different lists of positions. In fact, only 126 lists occur. This is a consequence of a result noticed by Kantor [15]: The symmetric differences in the symplectic designs $S(2m)$ correspond to the *hyperplanes* in the $2m$-dimensional discrete affine space on the finite field of order 2, denoted $\mathbb{F}_2$. In the case $m = 3$, there are 126 such hyperplanes, each containing $2^5 = 32$ points. Each hyperplane is the symmetric difference of 16 different choices of block pairs.

From Kantor's work, we can also extract a criterion for when a set of three blocks $\{B_i, B_j, B_k\}$ will be part of an S-incompatible quadruple. As we deduced, this occurs when the symmetric difference of the three blocks is the complement of a block. The quantity

$$\left| (B_i \ominus B_j) \cap B_k \right| \tag{5.85}$$

equals either 16 or 20, depending on the choice of blocks. When it equals 16, the symmetric difference of the three blocks is itself a block. On the other hand, when it equals 20, then the symmetric difference is the complement of a block, and we have the incompatibility we seek. This can be interpreted in terms of another affine space on the finite field $\mathbb{F}_2$. In this space, the points are the 64 bitstrings of the twin Hoggar SIC. For a fixed B_i and B_j with $j \neq i$, the set of all B_k such that

$$\left| (B_i \ominus B_j) \cap B_k \right| = 16 \tag{5.86}$$

defines a hyperplane in this affine space. Points that lie outside this hyperplane correspond to bitstrings which, together with B_i and B_j, can form part of an S-incompatible quartet.

This construction also tells us about the triple products, in a way that relates back to our symplectic bilinear form, Eq. (5.44). Consider the quartet formed by B_i, B_j, B_k and their symmetric difference. If this quartet is S-incompatible, then

$$\mathrm{Re\,tr}(\Pi_i \Pi_j \Pi_k) = \mathrm{Re\,tr}(\pi_i \pi_j \pi_k) = 0. \tag{5.87}$$

In dimension 3, the triple products of the Hesse SIC depend on whether or not three points are collinear [24]. Now, we see that in dimension 8, triple products depend upon whether three points lie in the same hyperplane.

In fact, the implication works both ways: If $(ijk) \in S_0$, then B_i, B_j and B_k can be extended to form an S-incompatible quartet.

5.7   From Pauli Operators to Real Equiangular Lines

We are now nicely positioned to revisit the connection between real and complex equiangular lines that we met in Sect. 6.4. That is, it's time to bring together $\mathbb{C}^8$ and $\mathbb{R}^7$. The bit $B_j(n)$ will be 0 if the inner product

$$\mathrm{tr}(\Pi_n \pi_j) = \mathrm{tr}(D_n \Pi_0 D_n^\dagger D_j \pi_0 D_j^\dagger) \tag{5.88}$$

vanishes. Here, the displacement operators D_n and D_j are built from tensor products of the Pauli matrices. Note that we can use the cyclic property of the trace to reduce the problem to investigating inner products of the form

$$\mathrm{tr}(\Pi_0 D_m \pi_0 D_m^\dagger) . \tag{5.89}$$

The product $\Pi_0 \pi_0$ is a symmetric matrix. If we want the trace to vanish, we should try introducing an asymmetry somehow.

Of the four Pauli matrices, three (counting the identity) are symmetric. Only Y, which is proportional to the product XZ, is antisymmetric. We therefore make the educated guess that the inner product will vanish if the displacement operator D_m involves *an odd number of factors* of the Pauli matrix Y. This happens in 28 out of the 64 possible displacement operators D_m, which is the number we're looking for. Why 28? If we want one factor of Y, we have three places to put it, and we have $3^2 = 9$ choices for the other two factors. This gives us 27 possible operators. Then, the operator YYY is also antisymmetric, making a total of 28.

It is straightforward to check that these zeros fall in the correct places to reproduce the first row of Table 6.1.

The displacement operator D_m will be antisymmetric if a certain sum has odd parity:

$$m_0 m_1 + m_2 m_3 + m_4 m_5 = 1 \quad \mathrm{mod}\ 2 . \tag{5.90}$$

This construction for picking 28 configurations out of 64 also arises in the study of *bitangents* to *quartic curves* [26]. Take the plane $\mathbb{R}^2$, and define a curve on the plane by a fourth-degree equation in two variables. Such a curve can have as many as 28 bitangent lines, i.e., lines that are tangent to the curve at exactly two places. By extending to the complex projective plane, one can always find a full set of 28 bitangents. Each one is labeled by a set of binary coordinates satisfying Eq. (5.90).

Rather unexpectedly, then, the study of SICs has made contact with the theory of algebraic curves!

Consider the elements of the index k that indicate the powers to which we raise X when constructing D_k, that is, the ordered triple (k_0, k_2, k_4). This triple can take eight different values, seven of them nonzero. Likewise, we have seven nonzero possibilities for (k_1, k_3, k_5). Let us group the possibilities for these two ordered triples according to when the dot product has even parity:

$$k_0 k_1 + k_2 k_3 + k_4 k_5 = 0 \quad \text{mod } 2 . \tag{5.91}$$

For each choice of (k_1, k_3, k_5), there are three choices for (k_0, k_2, k_4) that satisfy Eq. (5.91).

$$
\begin{array}{cc}
(k_0, k_2, k_4) & (k_1, k_3, k_5) \\
010, 011, 001 & 100 \\
001, 101, 100 & 010 \\
010, 110, 100 & 001 \\
001, 111, 110 & 110 \\
010, 111, 101 & 101 \\
011, 111, 100 & 011 \\
110, 101, 011 & 111
\end{array}
\tag{5.92}
$$

This configuration has a name: It is simply the Fano plane! The choices for (k_0, k_2, k_4) label the points, and and (k_1, k_3, k_5) label the lines. A line and a point are incident if and only if their coordinates satisfy Eq. (5.91).

In the Fano plane, there are 28 ways to select a point and a line *not* incident with it: For each point, four of the seven lines do not go through that point, and we have seven ways to choose a point. In discrete geometry, a *flag* is the combination of a line and a point lying on that line, and an *anti-flag* is a line with a point lying off that line. So, there are 28 anti-flags in the Fano plane, and for each of them, the dot product of the point and line labels has *odd* parity. That is, for each anti-flag, the label of the point and the label of the line satisfy Eq. (5.90).

Look back at Table 6.1. Each occurrence of the bit 0 is an anti-flag in a Fano plane! We use the powers to which we raise X to pick a point, and the powers to which we raise Z to pick a line (or vice versa). If the point lies off the line, we write a 0. All other bits in the sequence, we set to 1.

Recalling our survey of real equiangular lines back in Sect. 1.2, we see that the orthogonalities between a Hoggar vector and the 28 vectors in the "entropic dual" SIC correspond to a maximal set of equiangular lines in $\mathbb{R}^7$. We have here another unforeseen relation between the complex and the real versions of the equiangular lines question. Starting with one maximal set of complex equiangular lines, we construct another, living in a different dimension over a different number field!

5.8   Concluding Remarks

SICs are a confluence of multiple topics in mathematics. Weyl–Heisenberg SIC solutions in dimensions larger than 3 turn out to have deep number-theoretic properties, connecting quantum information theory to Hilbert's twelfth problem [27]. The other known SIC solutions, which we have termed the sporadic SICs, relate by way of group theory to sphere packing and the octonions [2]. By asking a physicist's question—"Given this constraint, which states maximize and minimize the entropy?"—we launched ourselves into symplectic designs, two-graphs and bitangents to quartic curves. In the next chapter, we will see how the exceptional Lie algebras also enter this picture. Prolonged exposure to the SIC problem makes one suspect that the interface between physics and mathematics does not have the shape that one first expected.

For each of the SICs with doubly transitive symmetry groups, the pure states that minimize the Shannon entropy of the SIC representation are related to equiangular real lines. In dimension 2, they form a SIC [12], which is a tetrahedron in the Bloch ball, and that yields four real lines. In dimension 3, they form 12 MUB states [24]. Picking one state from each MUB, we obtain four equiangular lines in nine-dimensional real space. (There are 81 ways to do this.) And in dimension 8, the procedure yields the twin Hoggar SIC, which is equivalent to 64 equiangular lines in $\mathbb{C}^8$. Furthermore, when we consider the relation between the original SIC and its twin, we find a set of 28 lines, which are a maximal set for 7- or 8-dimensional real vector space. And, as we remarked before, the triple-product structure of the Hoggar SIC leads to a two-graph on 64 vertices, which is itself equivalent to a set of equiangular lines in $\mathbb{R}^{36}$.

That the solutions to the real and complex versions of the equiangular lines problem should be related in this way is rather surprising.

To draw this essay to a close, we should note that the Hoggar SIC provides a rather clean and elementary introduction to several mathematical structures that have been employed in the study of three-qubit quantum systems [28]. For example, we encountered the group PSU(3, 3): It was (up to isomorphism) simply the group of transformations that permute the vectors in the Hoggar SIC while leaving the fiducial untouched. This group has also appeared [29] in studies of Bell–Kochen–Specker phenomena, that is, of the nonclassical meshing together of probability assignments [24, 30–33]. Likewise, the sorting of tensor products of Pauli operators into symmetric and antisymmetric matrices has been invoked in other problems [34]. In the next chapter, we will see how the Hoggar-type SICs lead to a polytope related to an exceptional Lie algebra; this, too, is a type of structure pertinent to Bell–Kochen–Specker phenomena in three-qubit systems [35, 36]. All this suggests that more ideas might yet be grown from the Hoggar SIC.

References

1. J.H. Conway, D. Smith, *On Quaternions and Octonions: Their Geometry, Arithmetic, and Symmetry* (A K Peters, Natick, 2003)
2. B.C. Stacey, Sporadic SICs and the normed division algebras. Found. Phys. **47**, 1060–64 (2017). https://doi.org/10.1007/s10701-017-0087-2
3. J.C. Baez, H. Joyce, Ubiquitous octonions (2005). https://plus.maths.org/content/ubiquitous-octonions
4. H. Zhu, Quantum state estimation and symmetric informationally complete POMs. Ph.D. thesis, National University of Singapore (2012). http://scholarbank.nus.edu.sg/bitstream/handle/10635/35247/ZhuHJthesis.pdf
5. H. Zhu, Super-symmetric informationally complete measurements. Ann. Phys. (NY) **362**, 311–326 (2015). https://doi.org/10.1016/j.aop.2015.08.005
6. R. Wilson et al., ATLAS of finite group representations (2016). http://brauer.maths.qmul.ac.uk/Atlas/v3/group/G22/
7. GAP – Groups, Algorithms, and Programming, version 4.8.3 (2016). http://www.gap-system.org
8. H.S. Ryser, Chapter 8: combinatorial designs, in *Combinatorial Mathematics* (Carus Monographs #14) (Mathematical Association of America, Washington, 1963)
9. D.M. Appleby, S.T. Flammia, C.A. Fuchs, The Lie algebraic significance of symmetric informationally complete measurements. J. Math. Phys. **52**, 022202 (2011). https://doi.org/10.1063/1.3555805
10. P.J. Cameron, J.H. van Lint, *Graphs, Codes and Designs* (Cambridge University Press, Cambridge, 1980)
11. D.E. Taylor, Two-graphs and doubly transitive groups. J. Comb. Theory, Ser. A **61**, 113–122 (1992). https://doi.org/10.1016/0097-3165(92)90056-Z
12. A. Szymusiak, W. Słomczyński, Informational power of the Hoggar symmetric informationally complete positive operator-valued measure. Phys. Rev. A **94**, 012122 (2015). https://doi.org/10.1103/PhysRevA.94.012122
13. D.M. Appleby, Å. Ericsson, C. Fuchs, Properties of QBist state spaces. Found. Phys. **41**, 564–579 (2011). https://doi.org/10.1007/s10701-010-9458-7
14. J.W. Iverson, D.G. Mixon, Doubly transitive lines II: almost simple symmetries (2019). arXiv:1905.06859
15. W.M. Kantor, Symplectic groups, symmetric designs, and line ovals. J. Algebra **33**, 43–58 (1975). https://doi.org/10.1016/0021-8693(75)90130-1
16. C. Parker, E. Spence, V.E. Tonchev, Designs with the symmetric difference property on 64 points and their groups. J. Comb. Theory, Ser. A **67**, 23–43 (1994). https://doi.org/10.1016/0097-3165(94)90002-7
17. P.J. Cameron, Square 2-designs, in *The Encyclopædia of Design Theory* (2003). http://www.maths.qmul.ac.uk/~leonard/designtheory.org/library/encyc/
18. J.F. Dillon, J.R. Schatz, Block designs with the symmetric difference property, in *Proceedings of the NSA Mathematical Sciences Meetings*. US Govt. Printing Office (1987). http://www.opensourcemath.org/papers/dillon-shatz-designs.pdf
19. C.M. Caves, C.A. Fuchs, R. Schack, Conditions for compatibility of quantum state assignments. Phys. Rev. A **66**(6) (2002). https://doi.org/10.1103/PhysRevA.66.062111
20. B. Allen, B.C. Stacey, Y. Bar-Yam, An information-theoretic formalism for multiscale structure in complex systems (2014). arXiv:1409.4708
21. B.C. Stacey, Multiscale structure in eco-evolutionary dynamics. Ph.D. thesis, Brandeis University (2015). arXiv:1509.02958
22. B.C. Stacey, B. Allen, Y. Bar-Yam, Multiscale information theory for complex systems: theory and applications, in *Information and Complexity*, eds. by M. Burgin, C.S. Calude (World Scientific, Singapore, 2017)
23. C.M. Caves, Symmetric informationally complete POVMs (2002). http://info.phys.unm.edu/~caves/reports/infopovm.pdf

24. B.C. Stacey, SIC-POVMs and compatibility among quantum states. Mathematics **4**(2), 36 (2016). https://doi.org/10.3390/math4020036
25. F. Ruskey, m. Weston, A survey of Venn diagrams. Electron. J. Comb. **DS#5**, n.p. (2005). http://www.combinatorics.org/files/Surveys/ds5/VennWhatEJC.html
26. J. Gray, From the history of a simple group. Math. Intell. **4**(2), 59–67 (1982). https://doi.org/10.1007/BF03023483
27. M. Appleby, S. Flammia, G. McConnell, J. Yard, Generating ray class fields of real quadratic fields via complex equiangular lines (2016). arXiv:1604.06098
28. B.L. Cerchiai, B. van Geemen, From qubits to E_7. J. Math. Phys. **51**, 12203 (2010). https://doi.org/10.1063/1.3519379
29. M. Planat, M. Saniga, F. Holweck, Distinguished three-qubit 'magicity' via automorphisms of the split Cayley hexagon. Quant. Info. Proc. **12**, 2535–49 (2013)
30. C.A. Fuchs, R. Schack, Quantum-Bayesian coherence. Rev. Mod. Phys. **85**, 1693–1715 (2013). https://doi.org/10.1103/RevModPhys.85.1693
31. C.A. Fuchs, B.C. Stacey, Some negative remarks on operational approaches to quantum theory, in *Quantum Theory: Informational Foundations and Foils*, eds. G. Chiribella, R.W. Spekkens (Springer, Berlin, 2016), pp. 283–305. https://doi.org/10.1007/978-94-017-7303-4_9
32. N.D. Mermin, Hidden variables and the two theorems of John Bell. Rev. Mod. Phys. **65**(3), 803–15 (1993). https://doi.org/10.1103/RevModPhys.65.803
33. N.D. Mermin, Erratum: Hidden variables and the two theorems of John Bell. Rev. Mod. Phys. **88**(3), 039902 (2016). https://doi.org/10.1103/RevModPhys.88.039902
34. P. Lévay, M.S., Vrana, P.: Three-qubit operators, the split Cayley hexagon of order two, and black holes. Phys. Rev. D **78**(12), 124002 (2008). https://doi.org/10.1103/PhysRevD.78.124022
35. M. Waegell, P.K. Aravind, Parity proofs of the Kochen-Specker theorem based on the Lie algebra E8. J. Phys. A **48**(22), 225301 (2015). https://doi.org/10.1088/1751-8113/48/22/225301
36. L. Loveridge, R. Dridi, The many mathematical faces of Mermin's proof of the Kochen–Specker theorem (2015). arXiv:1511.00950

Chapter 6
Sporadic SICs and the Exceptional Lie Algebras

*Sometimes, mathematical oddities crowd in upon one another,
and the exceptions to one classification scheme reveal
themselves as fellow-travelers with the exceptions to a quite
different taxonomy. After laying down some preliminaries, we
will establish a connection between the sporadic SICs and the
exceptional Lie algebras $\mathfrak{e}_6$, $\mathfrak{e}_7$ and $\mathfrak{e}_8$ by way of their root
systems.*

6.1 Root Systems and Lie Algebras

A *root system* is a set of vectors that can act as both "code" and "data" [1]. If we write
a vector $\mathbf{v}$, what instruction can we think of it as specifying? We can rotate around
it, but then we would also have to know how much to rotate. Also, we can reflect
through the plane orthogonal to it. Given $\mathbf{v}$, we can transform any other vector $\mathbf{u}$ to

$$\mathbf{u}' := \mathbf{u} - \frac{2(\mathbf{v} \cdot \mathbf{u})}{\mathbf{v} \cdot \mathbf{v}} \mathbf{v}. \tag{6.1}$$

Any time we define a map, we should ask what it preserves. Having defined a map
from vectors to vectors, are there sets of vectors that it leaves unchanged?

The most essential requirement for a root system is that the transformations defined
by the vectors in it leave the root system unchanged overall. They may permute the
set of vectors, but we neither gain nor lose vectors in the process. That is, a root
system is a finite set of vectors S in some vector space V, such that for each $\mathbf{u}, \mathbf{v} \in S$,
the new vector $\mathbf{u}'$ calculated as above also belongs to S. Without loss of generality,
we can say that S will span V. Evidently, S will span a subspace of V, and nothing
very interesting will happen outside of that subspace, so we might as well just ignore
it. We exclude the zero vector from S, because it generates no reflection. Likewise,

if $\mathbf{v} \in S$, then $-\mathbf{v}$ will also belong to V, but including any other scalar multiples of $\mathbf{v}$ will just give us the same reflections again, so to avoid redundancy we impose the condition that the only scalar multiple of $\mathbf{v}$ we allow is $-\mathbf{v}$. Finally, to trim down the possibilities, we insist that the quantity $2(\mathbf{v} \cdot \mathbf{u})/(\mathbf{v} \cdot \mathbf{v})$ that appeared in the reflection formula is always an *integer* for any $\mathbf{u}, \mathbf{v} \in S$.

We call the vectors of a root system the roots. The reflections defined by the roots generate a group of transformations known as the *Weyl group* of the root system.

One rather elementary example of a root system is the set of vectors from the origin to the vertices of a regular hexagon centered at the origin. The Weyl group of this root system is just the group of permutations of three objects. In this case, the pairs of opposite roots themselves comprise a set of equiangular lines; the relations we will see later between the two kinds of structure will be more subtle.

A good thing to know about any kind of structure is whether it can be dissassembled into smaller structures of the same type. What are the "prime numbers" of root systems, so to speak? Instead of calling root systems "composite", we say that a root system is *reducible* if it can be split into two collections such that every vector in one is orthogonal to every vector in the other. It turns out that every root system can be written as the combination of *irreducible* root systems in a unique way, up to reordering.

Years of experience with vectors suggests that some notion of a "basis" for a root system will be important. The idea that proves fruitful is to ask when a root can be written as a linear combination of other roots with integer coefficients. A *base* for a root system is a set of roots that span the space where the system lives, with the property that any root $\mathbf{v}$ can be expressed as a sum over the base roots,

$$\mathbf{v} = \sum_i \alpha_i \mathbf{b}_i , \tag{6.2}$$

and for a given $\mathbf{v}$ all of the coefficients $\{\alpha_i\}$ are either nonnegative or nonpositive integers.

Let $\mathbf{u}$ and $\mathbf{v}$ be roots in a root system S, and consider the product

$$\frac{2(\mathbf{v} \cdot \mathbf{u})}{\mathbf{v} \cdot \mathbf{v}} \frac{2(\mathbf{u} \cdot \mathbf{v})}{\mathbf{u} \cdot \mathbf{u}} = 4 \cos^2 \theta , \tag{6.3}$$

where θ is the angle between the two roots. The condition that the factors be integers implies that this product must be an integer, and we know it must be between 0 and 4. And the only way we can have $\cos^2 \theta = 1$ is if $\mathbf{u} = \pm \mathbf{v}$. Suddenly the classification of irreducible root systems is starting to look rather tractable!

The visual technology that we use to carry out that classification is known as *Dynkin diagrams*. A Dynkin diagram is a decorated graph: a set of nodes (circles) connected by lines. Each line has a value attached, typically represented by drawing it with single, double or triple thickness. Each node stands for a base root in a root system, and the thickness of the edge connecting two nodes denotes the angle between those roots. We can classify the irreducible root systems by finding the rules

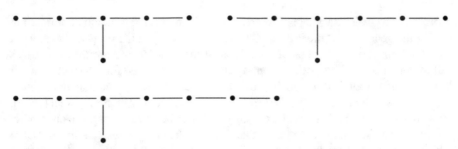

Fig. 6.1 Dynkin diagrams for those exceptional root systems that we will need: E_6 (top left), E_7 (top right) and E_8. The other two possibilities, denoted G_2 and F_4, involve edges with different thicknesses

by which a geometrically valid Dynkin diagram can be modified to give another, and identifying the patterns which cannot appear in any legitimate Dynkin diagram [2]. The result is that the allowed diagrams fall into several infinite families, along with a few exceptions which stand outside those families.

Perhaps the most significant role of root systems is in group theory. A *Lie group* is a group that is uncountably infinite, but in a manageable way: a group that is also a manifold. The *Lie algebra* of a Lie group expresses how that group looks near its identity element [3]. How such a thing might get boiled down to a root system is not at all obvious, and indeed, the standard story is a bit of an arduous one, full of steps that seem to lack motivation and technicalities that need to be squared away in order to ensure that the structure of one object carries over to the structure of another in a meaningful manner. A physics student can perhaps appreciate the big picture by thinking about angular momentum in quantum mechanics. Sometime during their first or second undergraduate encounter with quantum mechanics, they feel their way around "angular momentum eigenstates" that are labeled by "quantum numbers" and manipulated by "raising and lowering operators". A good course would connect angular momentum with spatial rotations, showing among other things that generators of rotations around different axes obey the commutator relation

$$[L_j, L_k] = i \sum_l \varepsilon_{jkl} L_l, \tag{6.4}$$

with ε_{jkl} being a completely antisymmetric object. Angular momentum, and the raising and lowering operators and all that, has to do with the symmetries of 3D space. Might then there be other kinds of spaces with other symmetries, for which moving between the analogues of angular momentum eigenstates involves a richer choice of vectors than just an up and a down arrow?

This, in an embarrassingly impressionistic way, is the story of how a Lie algebra gets its root system. One selects a representation of the algebra—a mapping of it into matrices of a chosen size—and from this extracts a set of vectors called *weights*. These are akin to the lists of "quantum numbers" that label the familiar angular momentum eigenstates; their components are eigenvalues of mutually commuting operators. The weights for a particularly important representation, which turn out to govern the generalized "raising" and "lowering", are called the roots and constitute a root system. The Lie groups of most interest to physicists are groups of matrices full of complex numbers. Pairs of opposing vectors $\pm\mathbf{v}$ in the root systems of their Lie algebras correspond to subalgebras that look just like the angular-momentum algebra we saw a moment ago. The full story is much longer than this précis [2–5]. However, most of the details of that story will not be of great consequence here; for our purposes, what will matter is that there are five *exceptional Lie algebras,* of which we will need the three largest. These algebras are described by the Dynkin diagrams in Fig. 6.1. We will try to stick to the convention that roman type indicates root systems and lattices while Fraktur type denotes Lie algebras, so the algebras described by these diagrams are $\mathfrak{e}_6$, $\mathfrak{e}_7$ and $\mathfrak{e}_8$.

6.2 E₆

In what follows, I will refer to H. S. M. Coxeter's [6]. Coxeter devotes a goodly portion of Chap. 12 to the *Hessian polyhedron*, which lives in $\mathbb{C}^3$ and has 27 vertices. These 27 vertices lie on nine diameters in sets of three apiece. (In a real vector space, only two vertices of a convex polyhedron can lie on a diameter. But in a complex vector space, where a diameter is a complex line through the center of the polyhedron, we can have more [7].) He calls the polyhedron "Hessian" because its nine diameters and twelve planes of symmetry interlock in a particular way. Their incidences reproduce the *Hesse configuration,* a set of nine points on twelve lines such that four lines pass through each point and three points lie on each line.

Coxeter writes the 27 vertices of the Hessian polyhedron explicitly, in the following way. First, let ω be a cube root of unity, $\omega = e^{2\pi i/3}$. Then, construct the complex vectors

$$(0, \omega^\mu, -\omega^\nu), \quad (-\omega^\nu, 0, \omega^\mu), \quad (\omega^\mu, -\omega^\nu, 0), \tag{6.5}$$

where μ and ν range over the values 0, 1 and 2. As Coxeter notes, we could just as well let μ and ν range over 1, 2 and 3. He prefers this latter choice, because it invites a nice notation: We can write the vectors above as

$$0\mu\nu, \ \nu0\mu, \ \mu\nu0. \tag{6.6}$$

For example,

$$230 = (\omega^2, -1, 0), \tag{6.7}$$

and

$$103 = (-\omega, 0, 1). \tag{6.8}$$

Coxeter then points out that this notation was first introduced by Beniamino Segre, "as a notation for the 27 lines on a general cubic surface in complex projective 3-

space. In that notation, two of the lines intersect if their symbols agree in just one place, but two of the lines are skew if their symbols agree in two places or nowhere." Consequently, the 27 vertices of the Hessian polyhedron correspond to the 27 lines on a cubic surface "in such a way that two of the lines are intersecting or skew according as the corresponding vertices are non-adjacent or adjacent."

Every smooth cubic surface in the complex projective space $\mathbb{C}P^3$ has exactly 27 lines that can be drawn on it. For a special case, the *Clebsch surface*, we can actually get a real surface (that is, in $\mathbb{R}P^3$) that we can mold in plaster and contemplate in the peaceful stillness of a good library. Intriguingly, the coefficients for the lines on the Clebsch surface live in the "golden field" $\mathbb{Q}(\sqrt{5})$, which we will meet again later in this chapter.

Casting the Hessian polyhedron into the real space $\mathbb{R}^6$, we obtain the polytope known as 2_{21}, which is related to $\mathfrak{e}_6$, since its symmetries (its "Coxeter group") are just the Weyl group of the E_6 root system. The notation here, due to Coxeter, refers to the lengths of the branches in the E_6 Dynkin diagram (Fig. 6.1), which in this context captures the reflection symmetries of the polytope. The Weyl group of E_6 can also be thought of as the Galois group of the 27 lines on a cubic surface.

We make the connection to symmetric quantum measurements by following the trick that Coxeter uses in his Eq. (12.39). We transition from the space $\mathbb{C}^3$ to the complex projective plane by collecting the 27 vertices into equivalence classes, which we can write in homogeneous coordinates as follows:

$$\begin{array}{lll} (0, 1, -1), & (-1, 0, 1), & (1, -1, 0) \\ (0, 1, -\omega), & (-\omega, 0, 1), & (1, -\omega, 0) \\ (0, 1, -\omega^2), & (-\omega^2, 0, 1), & (1, -\omega^2, 0) \end{array} \tag{6.9}$$

Let **u** and **v** be any two of these vectors. We find that

$$|\langle \mathbf{u}, \mathbf{u} \rangle|^2 = 4 \tag{6.10}$$

when the vectors coincide, and

$$|\langle \mathbf{u}, \mathbf{v} \rangle|^2 = 1 \tag{6.11}$$

when **u** and **v** are distinct. We can normalize these vectors to be quantum states on a three-dimensional Hilbert space by dividing each vector by $\sqrt{2}$.

We have found a SIC for $d = 3$. When properly normalized, Coxeter's vectors furnish a set of $d^2 = 9$ pure quantum states, such that the magnitude squared of the inner product between any two distinct states is $1/(d + 1) = 1/4$.

To relate the Coxeter construction to our way of generating SICs by group orbits, turn the first of Coxeter's vectors into a column vector:

$$\begin{pmatrix} 0 \\ 1 \\ -1 \end{pmatrix}. \tag{6.12}$$

Apply the cyclic shift operator X twice in succession to get the other two vectors in Coxeter's table (converted to column-vector format). Then, apply the phase operator Z twice in succession to recover the right-hand column of Coxeter's table. Finally, apply X to these vectors again to effect cyclic shifts and fill out the table. This gives us back the familiar Hesse SIC.

Each of the 27 lines corresponds to a "weight in the minimal representation" of $\mathfrak{e}_6$ [8]; "minimal" here means "of smallest dimensionality". (Roots and weights coincide in the case of $\mathfrak{e}_8$, where in the jargon, the adjoint representation is the minimal representation. We will talk more about $\mathfrak{e}_8$ in the next section.) And so, each element in the Hesse SIC corresponds to three weights of $\mathfrak{e}_6$.

Recall that starting with the Hesse SIC, we can obtain a full set of MUB by extremizing the Shannon entropy. Considering all the lines in the original structure that are orthogonal to a given line in its "entropic dual" SIC yields a maximal set of real equiangular lines in one fewer dimensions. (Oddly, I noticed this happening up in dimension 8 before I thought to check in dimension 3 [9], but we'll get to that soon.) To visualize the step from $\mathbb{C}^3$ to $\mathbb{R}^2$, we can use the Bloch sphere representation for two-dimensional quantum state space. Pick a state in the dual structure, i.e., one of the twelve MUB vectors. All the SIC vectors that are orthogonal to it must crowd into a 2-dimensional subspace. In other words, they all fit into a qubit-sized state space, and we can draw them on the Bloch sphere. When we do so, they are coplanar and lie at equal intervals around a great circle, a configuration sometimes called a *trine* [10]. This configuration is a maximal equiangular set of lines in the plane $\mathbb{R}^2$.

What happens if, starting with the Hesse SIC, you instead consider all the lines in the dual structure that are orthogonal to a given vector in the original? This yields a SIC in dimension 2. I don't know where in the literature that is written, but it feels like something Coxeter would have known.

Another path from the sporadic SICs to E_6 starts with the qubit SICs, i.e., regular tetrahedra inscribed in the Bloch sphere. Shrinking a tetrahedron, pulling its vertices closer to the origin, yields a type of quantum measurement (sometimes designated a SIM [11]) that has more intrinsic noise. Apparently, E_6 is part of the story of what happens when the noise level becomes maximal and the four outcomes of the measurement merge into a single degenerate case. This corresponds to a singularity in the space of all rotated and scaled tetrahedra centered at the origin. Resolving this singularity turns out to involve the Dynkin diagram of E_6: We invent a smooth manifold that maps to the space of tetrahedra, by a mapping that is one-to-one and onto everywhere except the origin. The pre-image of the origin in this smooth manifold is a set of six spheres, and two spheres intersect if and only if the corresponding vertices in the Dynkin diagram are connected [12].

6.3   E₈

On the fourteenth of March, 2016, Maryna Viazovska published a proof that the E_8 lattice is the best way to pack hyperspheres in eight dimensions [13]. I celebrated the third anniversary of this event by writing a guest post at the n-Category Café, explaining how this relates to another packing problem that seems quite different: how to fit as many equiangular lines as possible into the *complex* space $\mathbb{C}^8$. The answer to this puzzle is another example of a SIC.

In the previous section, we saw how to build SICs by starting with a fiducial vector and taking the orbit of that vector under the action of a group, turning one line into d^2. We said that the Weyl–Heisenberg group was the group we use in call cases but one. Now, we take on that exception. It will lead us to the exceptional root systems E_7 and E_8. Actually, it will be a bit easier to tackle the latter first. Whence E_8 in the world of SICs?

We saw how to generate the Hesse SIC by taking the orbit of a fiducial state under the action of the $d = 3$ Weyl–Heisenberg group. Next, we will do something similar in $d = 8$. We start by defining the two states

$$\left|\psi_0^{\pm}\right\rangle \propto (-1 \pm 2i, 1, 1, 1, 1, 1, 1, 1)^{\mathsf{T}}. \tag{6.13}$$

Here, we are taking the transpose to make our states column vectors, and we are leaving out the dull part, in which we normalize the states to satisfy

$$\left\langle\psi_0^{+}\middle|\psi_0^{+}\right\rangle = \left\langle\psi_0^{-}\middle|\psi_0^{-}\right\rangle = 1. \tag{6.14}$$

First, we focus on $\left|\psi_0^{+}\right\rangle$. To create a SIC from the fiducial vector $\left|\psi_0^{+}\right\rangle$, we take the set of Pauli matrices, including the identity as an honorary member: $\{I, \sigma_x, \sigma_y, \sigma_z\}$. We turn this set of four elements into a set of sixty-four elements by taking all tensor products of three elements. This creates the Pauli operators on three qubits. By computing the orbit of $\left|\psi_0^{+}\right\rangle$ under multiplication (equivalently, the orbit of $\Pi_0^{+} = \left|\psi_0^{+}\right\rangle\!\left\langle\psi_0^{+}\right|$ under conjugation), we find a set of 64 states that together form a SIC set.

The same construction works for the other choice of sign, $\left|\psi_0^{-}\right\rangle$, creating another SIC with the same symmetry group. Both of them are SICs of Hoggar type.

Now, we recall what we observed in the previous chapter. The fiducial stabilizer for a Hoggar-type SIC is isomorphic to the automorphism group of the octavians, which up to an overall scaling are also the points of the E_8 lattice. The 240 unit-norm octavians are the roots of the $\mathfrak{e}_8$ algebra, which are also the vertices of the polytope called 4_{21}.

Coxeter arrived at these 240 points in a different way, by extending his construction of the Hessian polyhedron to the *Witting polytope*. This polytope has 240 vertices and lives in 4-dimensional complex space. There are 27 edges at each vertex, and if we slice off the part of the polytope that is incident upon a vertex, we get a Hessian polyhedron. Reading each complex coordinate as two real coordinates maps the Witting polytope into eight-dimensional Euclidean space. And in that guise, its vertices give the root vectors of $\mathfrak{e}_8$.

Despite the media hype, there does not appear to be a "Theory of Everything" dwelling within E_8 [14]. But there *is* a particularly nice quantum measurement with E_8 hiding in its symmetries! Before moving on, we pause to note how peculiar it is

Table 6.1 Four of the states from the $\{\Pi_i^-\}$ Hoggar-type SIC, written in the probabilistic representation of three-qubit state space provided by the $\{\Pi_i^+\}$ SIC. Up to an overall normalization by $1/36$, these states are all binary sequences, i.e., they are uniform over their supports

```
11101110111000011110111011100001111011101110000100010001000110
11011101110100101101110111010010110111011101001000100010001011
10111011101101001011101110110100101110111011010001000100010010
01110111011110000111011101111000011101110111100010001000100001
```

that by trying to find a nice packing of complex unit vectors, we ended up talking about an optimal packing of Euclidean hyperspheres [13].

Now that we've met E_8, it's time to visit the root system we skipped: Where does E_7 fit in?

6.4 E7

With respect to the probabilistic representation furnished by the Π_0^+ SIC, the states of the Π_0^- SIC minimize the Shannon entropy, and vice versa [9, 15].

Recall that when we invented SICs for a single qubit, they were tetrahedra in the Bloch ball, and we could fit together two tetrahedral SICs such that each vector in one SIC was orthogonal (in the Bloch picture, antipodal) to exactly one vector in the other. The two Hoggar-type SICs made from the fiducial states Π_0^+ and Π_0^- satisfy the grown-up version of this relation: Each state in one is orthogonal to exactly twenty-eight states of the other (Table 6.1).

We saw that we can understand these orthogonalities as corresponding to the antisymmetric elements of the three-qubit Pauli group. Moreover, these 28 antisymmetric matrices correspond exactly to the 28 *bitangents of a quartic curve,* and lines in a maximal equiangular set in $\mathbb{R}^7$. We had a taste of how this connection works in the previous chapter. Given any smooth plane quartic defined over $\mathbb{C}$, we can draw exactly 28 lines that are tangent to it at two points [16]. Much like with the 27 lines on a cubic surface, we can organize the 28 bitangents using a polytope, in this case a polytope in $\mathbb{R}^7$ whose 56 vertices come in opposite pairs. These diameters are a set of equiangular lines that saturate the Gerzon bound.

To recap: Each of the 64 vectors (or, equivalently, projectors) in the Hoggar SIC is naturally labeled by a displacement operator, which up to an overall phase is the tensor product of three Pauli operators. Recall that we can write the Pauli operator σ_y as the product of σ_x and σ_z, up to a phase. Therefore, we can label each Hoggar-SIC vector by a pair of binary strings, each three bits in length. The bits indicate the power to which we raise the σ_x and σ_z generators on the respective qubits. The pair (010, 101), for example, means that on the three qubits, we act with σ_x on the second, and we act with σ_z on the first and third. Likewise, $(000, 111)$ stands for the displacement operator which has a factor of σ_z on each qubit and no factors of σ_x at all.

There is a natural mapping from pairs of this form to pairs of unit octonions. Simply turn each triplet of bits into an integer and pick the corresponding unit from the set $\{1, e_1, e_2, e_3, e_4, e_5, e_6, e_7\}$, where each of the e_j square to -1.

We can choose the labeling of the unit imaginary octonions so that the following nice property holds. Up to a sign, the product of two imaginary unit octonions is a third, whose index is the XOR of the indices of the units being multiplied. For example, in binary, $1 = 001$ and $4 = 100$; the XOR of these is $101 = 5$, and e_1 times e_4 is e_5. Translating the indices of the Cayley–Graves table in Eq. (1.3) verifies that, up to sign factors, XOR implements octonion multiplication.

What, then, is the meaning of those sign factors?

They are the means by which we flesh out the seven vectors we got from the Fano plane's incidence matrix to a full set of twenty-eight! (Van Lint and Seidel noted that the incidence matrix of the Fano plane could be augmented into a set that saturates the Gerzon bound [17, 18], but to my knowledge, extracting the necessary choices of sign from octonion multiplication is not reported in the literature.)

The exceptional Lie algebras enter the story because our equiangular lines in $\mathbb{R}^7$ are the diameters of the *Gosset polytope* 3_{21}. Again, the notation here refers to the lengths of the branches in a Dynkin diagram, this time for E_7 (Fig. 6.1). So, because we have made our way to the polytope 3_{21}, we have arrived at E_7. To quote a fascinating paper by Manivel [8],

> Gosset seems to have been the first, at the very beginning of the 20th century, to understand that the lines on the cubic surface can be interpreted as the vertices of a polytope, whose symmetry group is precisely the automorphism group of the configuration. Coxeter extended this observation to the 28 bitangents, and Todd to the 120 tritangent planes. Du Val and Coxeter provided systematic ways to construct the polytopes, which are denoted n_{21} for $n = 2, 3, 4$ and live in $n + 4$ dimensions. They have the characteristic property of being semiregular, which means that the automorphism group acts transitively on the vertices, and the faces are regular polytopes. In terms of Lie theory they are best understood as the polytopes in the weight lattices of the exceptional simple Lie algebras $\mathfrak{e}_{n+4}$, whose vertices are the weights of the minimal representations.

When we studied the Hesse SIC, we met the case $n = 2$ and $\mathfrak{e}_6$. The intricate orthogonalities between two conjugate SICs of Hoggar type have led us to the case $n = 3$ and $\mathfrak{e}_7$. As Manivel describes, the stabilizer of any of the 28 bitangents is the automorphism group of the 27 lines on a cubic surface, the Weyl group of E_6. So, we have a peculiar way to descend indirectly from $\mathbb{C}^8$ through $\mathbb{R}^7$ to $\mathbb{C}^3$: The orthogonalities between two conjugate Hoggar-type SICs lead us to a polytope, and its symmetries are tallied by the vectors which, read projectively, furnish the Hesse SIC.

6.5   The Regular Icosahedron and Real-Vector-Space Quantum Theory

In the previous sections, we uncovered correspondences between equiangular lines in $\mathbb{C}^3$ and $\mathbb{R}^2$, and between $\mathbb{C}^8$ and $\mathbb{R}^7$. It would be nice to have a connection like that between $\mathbb{C}^4$ and $\mathbb{R}^3$, but I have not found one yet. Instead, there is a slightly different relationship that brings $\mathbb{R}^3$ into the picture.

Suppose that, unaccountably, we wished to build the Hesse SIC, but in *real* vector space. What might this even mean? It would entail finding a fiducial vector and an appropriate group, closely analogous to the qutrit Weyl–Heisenberg group, such that the orbit of said fiducial is a maximal set of equiangular lines. How big would such a set of lines be? Recall that the Gerzon bound is d^2 for $\mathbb{C}^d$, but only $d(d+1)/2$ in $\mathbb{R}^d$. In both cases, this is essentially because those values are the dimensions of the appropriate operator spaces (symmetric for operators on $\mathbb{R}^d$, self-adjoint for operators on $\mathbb{C}^d$). It is not difficult to show that, if the Gerzon bound is attained, the magnitude of the inner product between the vectors is $1/\sqrt{d+1}$ in $\mathbb{C}^d$ and $1/\sqrt{d+2}$ in $\mathbb{R}^d$.

We are familiar with the complex case, in which we define a shift operator X and a phase operator Z that both have order d. A cyclic shift is nice and simple, so we'd like to keep that idea, but the only "phase" we have to work with is the choice of positive or negative sign. So, let us consider the operators

$$X = \begin{pmatrix} 0 & 0 & 1 \\ 1 & 0 & 0 \\ 0 & 1 & 0 \end{pmatrix}, \text{ and } Z = \begin{pmatrix} 1 & 0 & 0 \\ 0 & -1 & 0 \\ 0 & 0 & 1 \end{pmatrix}. \tag{6.15}$$

The shift operator X still satisfies $X^3 = I$, while for the phase operator Z, we now have $Z^2 = I$.

What group can we make from these operators? Note that

$$(ZX)^3 = -I, \tag{6.16}$$

and so

$$(-Z)^2 = X^3 = (-ZX)^3 = I, \tag{6.17}$$

meaning that the operators X and $-Z$ generate the *tetrahedral group,* so designated because it is isomorphic to the rotational symmetry group of a regular tetrahedron. Equivalently, we can use Z as a generator, since $-Z = (ZX)^3 Z$ by the above.

Now, we want to take the orbit of a vector under this group! But what vector? It should not be an eigenvector of X or of Z, for then we know we could never get a full set. Therefore, we don't want a flat vector, nor do we want any of the basis vectors, so we go for the next simplest thing:

$$\mathbf{v} = \begin{pmatrix} 0 \\ 1 \\ y \end{pmatrix}, \tag{6.18}$$

where y is a real number. We now have

$$Z\mathbf{v} = \begin{pmatrix} 0 \\ -1 \\ y \end{pmatrix} \tag{6.19}$$

and also

$$X^2\mathbf{v} = \begin{pmatrix} 1 \\ y \\ 0 \end{pmatrix}, \tag{6.20}$$

so if we want equality between the inner products,

$$\langle Z\mathbf{v}, \mathbf{v} \rangle = \langle X^2\mathbf{v}, \mathbf{v} \rangle, \tag{6.21}$$

then we need to have

$$-1 + y^2 = y. \tag{6.22}$$

The positive solution to this quadratic equation is

$$y = \frac{1 + \sqrt{5}}{2}, \tag{6.23}$$

so we can in fact take our y to be ϕ, the golden ratio.

In the group we defined above, X performs cyclic shifts, Z changes the relative phase of the components, and we have the freedom to flip all the signs. Therefore, the orbit of the fiducial $\mathbf{v}$ is the set of twelve vectors

$$\begin{pmatrix} 0 \\ \pm 1 \\ \pm\phi \end{pmatrix}, \begin{pmatrix} \pm 1 \\ \pm\phi \\ 0 \end{pmatrix}, \begin{pmatrix} \pm\phi \\ 0 \\ \pm 1 \end{pmatrix}. \tag{6.24}$$

These are the vertices of a regular icosahedron, and the diagonals of that icosahedron are six equiangular lines. The inner products between these vectors are always $\pm\phi$. Since $d(d + 1)/2 = 6$, there cannot be any larger set of equiangular lines in $\mathbb{R}^3$.

Recall that the reciprocal of the golden ratio ϕ is

$$\phi^{-1} = \frac{2}{1 + \sqrt{5}} \frac{1 - \sqrt{5}}{1 - \sqrt{5}} = \frac{2 - 2\sqrt{5}}{1 - 5} = \frac{-1 + \sqrt{5}}{2} = \phi - 1. \tag{6.25}$$

The golden ratio ϕ is a root of the monic polynomial $y^2 - y - 1$, and being a root of a monic polynomial with integer coefficients, it is consequently an algebraic integer. The same holds for its reciprocal, so ϕ^{-1} is also an algebraic integer, making the two of them *units* in the number field $\mathbb{Q}(\sqrt{5})$'s ring of algebraic integers.

To summarize: For the diagonals of the regular icosahedron, the vector components are given by the units of the integer ring of the "golden field" $\mathbb{Q}(\sqrt{5})$. But it has been discovered [19] that the vector components for the Weyl–Heisenberg SICs in dimension *four* are derived from an integer-ring unit in the *ray class field over* $\mathbb{Q}(\sqrt{5})$. Therefore, in a suitably perplexing way, the icosahedron is the Euclidean version of the Hesse SIC, and the SICs in $d = 4$ are the number-theoretic extension of the icosahedron.

Futhermore, it has been observed empirically that in dimensions

$$d_k = \phi^{2k} + \phi^{-2k} + 1, \qquad (6.26)$$

there exist Weyl–Heisenberg SICs with additional group-theoretic properties that make their exact expressions easier to find. These are known as *Fibonacci–Lucas SICs* [20].

There are exactly four known cases where the Gerzon bound can be attained in $\mathbb{R}^d$: when $d = 2, 3, 7$ and 23. Three out of these four examples relate to SICs, specifically to the sporadic SICs. We can obtain the maximal equiangular sets in $\mathbb{R}^2$ and $\mathbb{R}^7$ from SICs in $\mathbb{C}^3$ and $\mathbb{C}^8$ respectively, while the set in $\mathbb{R}^3$ turns out to be the real analogue of our example in $\mathbb{C}^3$. All of this raises a natural question: What about $\mathbb{R}^{23}$? Does the equiangular set there descend from a SIC in $\mathbb{C}^{24}$? That, nobody knows.

Zhu has proved that there are no more doubly-transitive SICs above $d = 8$ [21]. So, there cannot be another sporadic SIC in $d = 24$ that is just like the qubit, Hesse and Hoggar structures, but this leaves plenty of room open for a SIC to be unusual in some other way.

We do know that the maximal equiangular line set in $\mathbb{R}^{23}$ can be extracted from the Leech lattice [22]. It contains 276 lines, and its automorphism group is Conway's group Co_3 [23]. Further study of this structure connects back with our use of SICs to give a probabilistic representation of quantum state space. Recall that when we fix a SIC in $\mathbb{C}^d$ as a reference measurement, we have a bound on the number of zero-valued entries within the probabilistic representation of a quantum state. We can also deduce the corresponding bound for the case of real equiangular lines. In $\mathbb{R}^d$, the Gerzon bound is $d(d + 1)/2$, and in those cases where we have a complete set of equiangular lines, we can play the game of doing "real-vector-space quantum mechanics", using our $d(d + 1)/2$ equiangular lines to define a reference measurement. As in $\mathbb{C}^d$, the Cauchy–Schwarz inequality gives an upper bound on the number of zero-valued entries in a probability distribution p, which works out to be

$$N_Z = \frac{d^2 - 1}{3}. \qquad (6.27)$$

Thinking about it for a moment, we realize that this is telling us about the maximum number of equiangular lines that can all simultaneously be orthogonal to a common vector. In particular, if we fix $d = 23$, we have a "real-vector-space SIC" that we can derive from the Leech lattice, and we know that

$$N_Z = \frac{23^2 - 1}{3} = 176 \tag{6.28}$$

of the elements of that set can be orthogonal to a "pure quantum state", i.e., a vector in $\mathbb{R}^{23}$. All 176 such lines have to crowd together into a 22-dimensional subspace, while still being equiangular. They comprise a maximal set of equiangular lines in $\mathbb{R}^{22}$, whose symmetries form the Higman–Sims finite simple group [24].

And that's what the classification of finite simple groups has to do with biodiversity!

6.6   Open Puzzles Concerning Exceptional Objects

While we're thinking about equiangular lines in spaces other than $\mathbb{C}^d$, here is a puzzle: What about the octonionic space $\mathbb{O}^3$, which figures largely in the study of exceptional objects? The Gerzon bound for this space works out to be 27. Cohn, Kumar and Minton give a nonconstructive proof that a set saturating the Gerzon bound in $\mathbb{O}^3$ exists, along with a numerical solution [25], but that numerical solution doesn't look like an approximation of a really pretty exact solution in any obvious way. (Their set of mutually unbiased bases in $\mathbb{O}^3$ *does* look like a generalization of a familiar set thereof in $\mathbb{C}^3$, which might raise our hopes.) Both in $\mathbb{R}^3$ and in $\mathbb{C}^3$, we can construct a maximal set of equiangular lines by starting with a fairly nice fiducial vector and applying a straightforward set of transformations. Is the analogous statement true in $\mathbb{O}^3$?

An equiangular set of 27 lines would provide a map from the set of density matrices for an "octonionic qutrit" to the probability simplex in $\mathbb{R}^{27}$, yielding a convex body that would be a higher-dimensional analogue of the Bloch ball. The extreme points of this Bloch body, the images of the "pure states", might form an interesting variety.

The 27 we have quoted here is related to a 27 that we encountered above. The algebra of self-adjoint operators on $\mathbb{O}^3$—the "observables" for an octonionic qutrit—is known as the *octonionic Albert algebra,* and it is 27-dimensional. The group of linear isomorphisms of $\mathbb{O}^3$ that preserve the determinant in the octonionic Albert algebra is a noncompact real form of E_6 [26]. As we saw earlier, the weights of the minimal representation of the Lie algebra $\mathfrak{e}_6$ yield the polytope 2_{21}, from which we can derive the Hesse SIC in $\mathbb{C}^3$. An exact solution for an "octonionic qutrit SIC" might close this circuit of ideas.

We met the Leech lattice back in Chap. 1. It lives in $\mathbb{R}^{24}$; one way to get at it is through $\mathbb{O}^3$ [27]. This suggests regarding the projections onto the 196,560 shortest nonzero vectors of the Leech lattice as states for an octonionic qutrit. It is also possible to fit the Leech lattice into the traceless part of the octonionic Albert algebra [28]. This means that each point in the Leech lattice is a "Hamiltonian" for a three-level octonionic quantum system. Moreover, recently it has been shown that the Standard Model gauge group can be extracted from the symmetries of an octonionic qutrit [29]. A better understanding of the structures which that state space supports might help clarify how much of that connection is coincidence and how much is physics.

Having reached a point where the tone has taken a rather speculative turn, we now embrace that attitude, just for the fun of it.

Another "28" that appears in the study of exceptional or unusual mathematical objects is the size of the *28-element Dedekind lattice*. This is a lattice in the sense of order theory, a partially ordered set with the property that we can trace subsets of elements upward through the ordering to where they join and downward to where they meet. It is the *free modular lattice on three generators* with the top and bottom elements removed, and Dedekind showed how to construct it as a sublattice within the lattice of subspaces of $\mathbb{R}^8$. Baez has suggested that its size is therefore related to the Lie group $SO(8)$, which is 28-dimensional [30]. Without resolving this conjecture, we note that the structure does sound a bit like something one would see in quantum theory, or a close relative of it. The "quantum logic" people have argued for a good long while that the lattice of closed subspaces of a Hilbert space, ordered by inclusion, can be thought of as a lattice of propositions pertaining to a quantum system. If the system in question is a set of three qubits, then we'd be talking about the lattice of subspaces of $\mathbb{C}^8$. To make this look exactly like the setting of Dedekind's lattice, we would have to do quantum mechanics over real vector space ("rebits" instead of qubits), but that's not so bad as far as pure math is concerned [31–33].

There's an idea, going back to Birkhoff and von Neumann in the 1930s, that in quantum physics, we should relax the distributive law of logic. The argument goes that we can measure the position of a particle, say, *or* we can measure its momentum, but per the uncertainty principle, we cannot precisely measure its position *and* its momentum at once. Thus, we should reconsider how the logical connectives

$$\wedge = \text{"and"}, \quad \vee = \text{"or"} \tag{6.29}$$

interact. I'm not convinced this is really the way to dig deep into the quantum mysteries: We can always impose an uncertainty principle on top of a classical theory [34, 35]. Still, the idea is good enough to wring some mathematics out of. In Boolean logic, we can distribute "or" over "and":

$$a \vee (b \wedge c) = (a \vee b) \wedge (a \vee c). \tag{6.30}$$

But if a, b and c are propositions about the outcomes of experiments upon a quantum system, then we cannot combine them arbitrarily and still have the result be physically meaningful. "Complementary" actions are mutually exclusive. We should only require that the distribution trick above works in restricted circumstances. When we organize all the propositions pertaining to a quantum system into a lattice, we say that

$a \leq b$ when a implies b. Then, unlike Boolean logic, we say that the distributive trick works when $a \leq b$ or $a \leq c$. This makes our lattice *modular* instead of *distributive*.

The free modular lattice on 3 generators is the structure we get when we introduce a set of elements $\{a, b, c\}$ and build up by combining them, assuming that the only restrictions are those due to requiring that the lattice be modular. In other words, it is the logic we build by starting with three propositions and saying nothing about them except that they should be "quantum" propositions, in this very stripped-down way of being "quantum".

We can also try approaching from the $SO(8)$ direction. Manogue and Schray point out that we can label a set of 28 generators for $SO(8)$ in the following way [36]. 7 of them correspond to the unit imaginary octonions e_1 through e_7. Then, for each of the e_i, there are three pairs of unit imaginary octonions (e_j, e_k) that multiply to e_i. These pairs give the other 21 generators. Considering the Fano plane, we have four generators for each point: one for the point itself, and one for each line through that point.

We can associate each generator with a line in $\mathbb{R}^7$ in the following way. First, we label each point in the Fano plane by the lines which meet there. For example,

$$(1, 1, 1, 0, 0, 0, 0) \tag{6.31}$$

stands for the point at which the first three lines coincide. There are seven such vectors, any two of which coincide at a single entry (because any two points in the Fano plane have exactly one line between them), and so any two of these vectors have the same inner product. Next, we pick one of the three lines through our chosen point, and we mark it by flipping the sign of that entry. For example,

$$(1, -1, 1, 0, 0, 0, 0) . \tag{6.32}$$

There are three ways to do this for each point in the Fano plane, and so we get 28 vectors in all. Note that the magnitude of the inner product is constant for all pairs. If the vectors correspond to different Fano points, then their supports overlap at only a single entry, and so the inner product is ± 1. If they correspond to the same Fano point, then the magnitude of the inner product is $1 - 1 + 1 = 1$ again.

So, the 28 of $SO(8)$ seems to be tied in with the other 28's: The same procedure counts generators of the group and equiangular lines in $\mathbb{R}^7$. (Counting—a kind of math I understand, some of the time.)

We now recall that the stabilizer subgroup for each vector in a Hoggar-type SIC is isomorphic to the projective special unitary group of 3×3 matrices over the finite field of order 9, known for short as PSU(3, 3). Finite-group theorists also refer to this structure as $U_3(3)$, and as $G_2(2)'$, since it is isomorphic to the commutator subgroup of the automorphism group of the *octavians,* the integer octonions. The symbol $G_2(2)'$ arises because the automorphism group of the octonions is called G_2, when we focus on the octavians we add a 2 in parentheses, and when we form the subgroup of all the commutators, we affix a prime.

We also recall that, given one SIC of Hoggar type, we can construct another by antiunitary conjugation, and each vector in the first SIC will be orthogonal to exactly 28 vectors out of the 64 in the other SIC. Furthermore, the Hilbert–Schmidt inner products that are not zero are all equal. Said another way, if we use the first SIC to define a probabilistic representation of three-qubit state space, then each vector in the second SIC is a probability distribution that is uniform across its support. Up to normalization, such a probability distribution is a binary string composed of 28 zeros and 36 ones.

Let $\{\Pi_j^+\}$ be a SIC of Hoggar type, and let $\{\Pi_j^-\}$ be its conjugate SIC. Suppose that U is a unitary that permutes the $\{\Pi_j^+\}$. Then a linear combination of 36 equally weighted projectors drawn from $\{\Pi_j^+\}$ will be sent to a linear combination of 36 equally weighted projectors from the set $\{\Pi_j^+\}$, possibly a different combination. But the only sequences of 36 ones and 28 zeros that correspond to valid quantum states are the representations of $\{\Pi_j^-\}$. Therefore, a unitary that shuffles the $+$ SIC will also shuffle the $-$ SIC. Furthermore, a unitary that *stabilizes* a projector, say Π_0^-, must permute the $\{\Pi_j^+\}$ in such a way that 1's go to 1's and 0's go to 0's.

To repeat: Because each SIC provides a basis, we can uniquely specify a vector in one SIC by listing the vectors in the other SIC with which it has nonzero overlap.

A unitary symmetry of one SIC set corresponds to a permutation of the other. Using the second SIC to define a representation of the state space, each vector in the first SIC is essentially a binary string, and sending one vector to another permutes the 1's and 0's. In particular, a unitary that stabilizes a vector in one SIC must permute the vectors of the second SIC in such a way that the list of 1's and 0's remains the same. The 1's can be permuted among themselves, and so can the 0's, but the binary sequence as a whole does not change.

It would be nice to have a way of visualizing this with a more tangible structure than eight-dimensional complex Hilbert space—something like a graph. Thinking about the permutations of 36 vectors, we imagine a graph on 36 vertices, and we try to draw it in such a way that its group of symmetries is isomorphic to the stabilizer group of a Hoggar fiducial. Can this be done? Well, almost—that is, up to a "factor of two": It's called the $U_3(3)$ graph, and its automorphism group has the stabilizer group of a Hoggar fiducial as an index-2 subgroup.

Now, is there a way to illustrate the structure of both SICs together as a graph? We want to record the fact that each vector in one SIC is nonorthogonal to exactly 36 vectors of the other, that the stabilizer of each vector is $PSU(3, 3)$, and that a stabilizer unitary shuffles the nonorthogonal set within itself. So, we start with one vertex to represent a fiducial vector, then we add 63 more vertices to stand for the other vectors in the first SIC, and then we add 36 vertices to represent the vectors in the second SIC that are nonorthogonal to the fiducial of the first. We'd like to connect the vertices in such a way that the stabilizer of any vertex is isomorphic to $PSU(3, 3)$. In fact, because any vector in either SIC can be identified by the list of the 36 nonorthogonal vectors in the other, the graph should look locally like the $U_3(3)$ graph everywhere! Is this possible? Yes! The result is the *Hall–Janko graph,* whose automorphism group has the Hall–Janko finite simple group as an index-2 subgroup.

The Hall–Janko group can also be constructed as the symmetries of the Leech lattice cast into a quaternionic form [37, 38]. Speculating only a little wildly, we can contemplate a possible path from the Hoggar-type SICs in $\mathbb{C}^8$ to the Hall–Janko group and thence to the Leech lattice and the "real SIC" in $\mathbb{R}^{23}$.

That's what we get when we think about the permutations of the 1's. What about PSU(3, 3) acting to permute the 0's?

This seems to lead us in the direction of the *Rudvalis group*. Wilson's textbook has this to say (Sect. 5.9.3):

> The Rudvalis group has order 145 926 144 000 $= 2^{14}.3^3.5^3.7.13.29$, and its smallest representations have degree 28. These are actually representations of the double cover 2 · Ru over the complex numbers, or over any field of odd characteristic containing a square root of -1, but they also give rise to representations of the simple group Ru over the field $\mathbb{F}_2$ of order 2.

Wilson then describes in some detail the 28-dimensional complex representations of 2 · Ru, using a basis in which $2^6 \cdot G_2(2)$ appears as the monomial subgroup [27]. And the 28 appears to be the same 28 we saw before; that is, it's the 28 pairs of cube roots of the identity in the octavians mod 2, so it's a set of 28 that is naturally shuffled by $G_2(2)'$ [39].

The Rudvalis group can be constructed as a rank-3 permutation group acting on 4060 points, where the stabilizer of a point is the Ree group $^2F_4(2)$. This group is in turn given by the symmetries of a "generalized octagon" [40]. (Generalized polygons abstract the properties of the more familiar polygons. In an ordinary polygon, each vertex is incident with two edges, each edge is incident with two vertices, and the circuit through the adjacent vertices and edges contains n of each. A generalized polygon replaces "two" with a larger integer. For example, the Fano plane is a generalized triangle.) Compare this situation to the Hall–Janko group, which has a rank-3 permutation representation on 100 points where the point stabilizer is $U_3(3)$, and $U_3(3)$ is furnished by the symmetries of a generalized hexagon [41].

Ru has a maximal subgroup given by a semidirect product $2^6 : G_2(2)' : 2$. This is what first caught my eye. Neglecting the issue of the group actions required to define the semidirect products, consider the factors: We have $G_2(2)'$ from the Hoggar stabilizer, 2 from conjugation and 2^6 from the three-qubit Pauli group.

This is reminiscent of a theorem proved by O'Nan [42]:

> Let G be a finite simple group having an elementary abelian subgroup E of order 64 such that E is a Sylow 2-subgroup of the centralizer of E in G and the quotient of the normalizer of E in G by the centralizer is isomorphic to the group $G_2(2)$ or its commutator subgroup $G_2(2)'$. Then G is isomorphic to the Rudvalis group.

The peculiar thing is that the Hall–Janko group is part of the "Happy Family", i.e., it is a subquotient of the Monster, while the Rudvalis group is a "pariah", floating off to the side. The Hoggar SIC almost seems to be acting as an intermediary between the two finite simple groups, one of which fits within the Monster while the other does not.

Finally, what connects the largest of sporadic simple groups with the second-smallest among quantum systems?

I was greatly amused to find the finite affine plane on nine points also appearing in the theory of the Monster group and the Moonshine module [43, 44]. In that case, the 9 points and 12 lines correspond to involution automorphisms. All the point-involutions commute with one another, and all the line-involutions commute with each other as well. The order of the product of a line-involution and a point-involution depends on whether the line and the point are incident or not.

This is probably of no great consequence—just an accident of the same small structures appearing in different places, because there are only so many small structures to go around. But it's a cute accident all the same.

References

1. J. Armstrong, Root systems recap (2010). https://unapologetic.wordpress.com/2010/03/12/root-systems-recap/
2. H. Georgi, *Lie Algebras in Particle Physics* (Westview Press, Boulder, 1999)
3. B.C. Hall, *Lie Groups, Lie Algebras, and Representations: An Elementary Introduction* (Springer, Berlin, 2003)
4. B. Salzberg, Buildings and shadows. Aequationes Mathematicae **25**, 1–20 (1982). https://doi.org/10.1007/BF02189594
5. S. Garibaldi, E_8, the most exceptional group. Bull. Am. Math. Soc. **53**, 643–71 (2016). https://doi.org/10.1090/bull/1540
6. H.S.M. Coxeter, *Regular Complex Polytopes*, 2nd edn. (Cambridge University Press, Cambridge, 1991)
7. H.S.M. Coxeter, The equiharmonic surface and the Hessian polyhedron. Annali di Matematica Pura ed Applicata **98**(1), 77–92 (1974). https://doi.org/10.1007/BF02414014
8. L. Manivel, Configurations of lines and models of Lie algebras. J. Algebra **304**(1), 457–86 (2006). https://doi.org/10.1016/j.jalgebra.2006.04.029
9. B.C. Stacey, Geometric and information-theoretic properties of the Hoggar lines (2016). arXiv:1609.03075
10. C.M. Caves, C.A. Fuchs, K.K. Manne, J.M. Renes, Gleason-type derivations of the quantum probability rule for generalized measurements. Found. Phys. **34**, 193–209 (2004). https://doi.org/10.1023/B:FOOP.0000019581.00318.a5
11. M.A. Graydon, D.M. Appleby, Quantum conical designs. J. Phys. A **49**, 085301 (2016). https://doi.org/10.1088/1751-8113/49/8/085301
12. J.C. Baez, The geometric McKay correspondence (part 1) (2017). https://golem.ph.utexas.edu/category/2017/06/the_geometric_mckay_correspond.html
13. M. Viazovska, The sphere packing problem in dimension 8. Ann. Math. **185**(3), 991–1015 (2017)
14. J. Distler, S. Garibaldi, There is no 'theory of everything' inside E_8. Commun. Math. Phys. **298**, 419–36 (2010). https://doi.org/10.1007/s00220-010-1006-y
15. A. Szymusiak, W. Słomczyński, Informational power of the Hoggar symmetric informationally complete positive operator-valued measure. Phys. Rev. A **94**, 012122 (2015). https://doi.org/10.1103/PhysRevA.94.012122
16. R.M. Green, *Combinatorics of Minuscule Representations* (Cambridge University Press, Cambridge, 2013)
17. J.H. van Lint, J.J. Seidel, Equiangular point sets in elliptic geometry. Proc. Nederl Akad. Wetensch Ser. A **69**, 335–48 (1966)
18. J.J. Seidel, A. Blokhuis, H.A. Wilbrink, J.P. Boly, C.P.M. van Hoesel, Graphs and association schemes, algebra and geometry. Technical report EUT-Report Vol. 83-WSK-02, Technische Hogeschool Eindhoven (1983). https://pure.tue.nl/ws/portalfiles/portal/1994552/253169.pdf

19. I. Bengtsson, The number behind the simplest SIC-POVM. Found. Phys. **47**, 1031–41 (2017). https://doi.org/10.1007/s10701-017-0078-3
20. M. Grassl, A.J. Scott, Fibonacci-Lucas SIC-POVMs. J. Math. Phys. **58**, 122201 (2017). https://doi.org/10.1063/1.4995444
21. H. Zhu, Super-symmetric informationally complete measurements. Ann. Phys. (NY) **362**, 311–326 (2015). https://doi.org/10.1016/j.aop.2015.08.005
22. H. Cohn, A. Kumar, Universally optimal distribution of points on spheres. J. Am. Math. Soc. **20**(1), 99–148 (2007). https://doi.org/10.1090/S0894-0347-06-00546-7
23. J.M. Goethals, J.J. Seidel, The regular two-graph on 276 vertices. Discret. Math. **12**, 143–158 (1975). https://doi.org/10.1016/B978-0-12-189420-7.50019-0. https://core.ac.uk/download/pdf/81988664.pdf
24. P.W.H. Lemmens, J.J. Seidel, Equiangular lines. J. Algebra **24**(3), 494–512 (1973)
25. H. Cohn, A. Kumar, G. Minton, Optimal simplices and codes in projective spaces. Geom. Topol. **20**(3), 1289–1357 (2016). https://doi.org/10.2140/gt.2016.20.1289
26. J.C. Baez, The octonions. Bull. AMS **39**, 145–205 (2002). https://doi.org/10.1090/S0273-0979-01-00934-X. http://math.ucr.edu/home/baez/octonions/octonions.html
27. R.A. Wilson, *The Finite Simple Groups* (Springer, Berlin, 2009)
28. J.C. Baez, G. Egan, Integral octonions (part 9) (2014). http://math.ucr.edu/home/baez/octonions/integers/integers_9.html
29. J.C. Baez, Octonions and the standard model (2021). https://math.ucr.edu/home/baez/standard/
30. J.C. Baez, Free modular lattice on 3 generators (2016). https://blogs.ams.org/visualinsight/2016/01/01/free-modular-lattice-on-3-generators/
31. L. Hardy, W.K. Wootters, Limited holism and real-vector-space quantum theory. Found. Phys. **42**, 454 (2012). https://doi.org/10.1007/s10701-011-9616-6
32. A. Aleksandrova, V. Borish, W.K. Wootters, Real-vector-space quantum theory with a universal quantum bit. Phys. Rev. A **87**, 052106 (2013). https://doi.org/10.1103/PhysRevA.87.052106
33. W.K. Wootters, Optimal information transfer and real-vector-space quantum theory (2013). arXiv:1301.2018
34. R.W. Spekkens, Evidence for the epistemic view of quantum states: a toy theory. Phys. Rev. A **75**(3), 032110 (2007). https://doi.org/10.1103/PhysRevA.75.032110
35. R.W. Spekkens, Reassessing claims of nonclassicality for quantum interference phenomena (2016). http://pirsa.org/16060102/
36. C.A. Manogue, J. Schray, Finite Lorentz transformations, automorphisms, and division algebras. J. Math. Phys. **34**, 3746–67 (1993). https://doi.org/10.1063/1.530056
37. J. Tits, Quaternions over $\mathbb{Q}(\sqrt{5})$, Leech's lattice and the sporadic group of Hall-Janko. J. Algebra **63**, 56–75 (1980). https://doi.org/10.1016/0021-8693(80)90025-3
38. A.M. Cohen, Finite quaternionic reflection groups. J. Algebra **64**, 293–324 (1980). https://doi.org/10.1016/0021-8693(80)90148-9. https://pure.tue.nl/ws/files/2467132/588246.pdf
39. J.F. Duncan, Vertex operators and sporadic groups, in *Moonshine — The First Quarter Century and Beyond: Proceedings of a Workshop on the Moonshine Conjectures and Vertex Algebras* (Cambridge University Press, Cambridge, 2010)
40. R.A. Wilson, A simple construction of the Ree groups of type $^2F_4$. J. Algebra **323**, 1468–81 (2010). https://doi.org/10.1016/j.jalgebra.2009.11.015
41. A.E. Schroth, How to draw a hexagon. Discret. Math. **199**, 161–171 (1999). https://doi.org/10.1016/S0012-365X(98)00294-5. https://core.ac.uk/download/pdf/82367708.pdf
42. M.E. O'Nan, A characterization of the Rudvalis group. Commun. Algebra **6**(2), 107–47 (1978). https://doi.org/10.1080/00927877808822236
43. M. Miyamoto, 21 involutions acting on the moonshine module. J. Algebra **175**, 941–65 (1995). https://doi.org/10.1006/jabr.1995.1220
44. T. Gannon, Monstrous moonshine: the first twenty-five years (2004). arXiv:math/0402345

Chapter 7
Exercises

We present a collection of opportunities for the reader to start making the SIC question their own.

The following problems are distilled from sources like research papers. Some may in practice require computer work doable with SymPy or similar tools. Exercises marked with a star have the potential to be more difficult than the others, not necessarily in terms of how much calculation needs to be done but on account of requiring more mental exploration. Hints to several problems may be found by combing the literature, although the intermediate steps might not be any more spelled out there than they are here. The number of problems included here is to be dismissed as a coincidence.

1. **Proving the Gerzon bound.** Let $\{\Pi_j : j = 1, \ldots, N\}$ be a set of projectors onto equiangular lines, so that $\mathrm{tr}\,\Pi_j \Pi_k = \alpha$ whenever $j \neq k$ and $\mathrm{tr}\,\Pi_j = 1$. Use the fact that these operators must be linearly independent to prove that N cannot exceed $d(d+1)/2$ when the lines live in $\mathbb{R}^d$ or d^2 when they live in $\mathbb{C}^d$. Prove that $\alpha = 1/(d+2)$ for $\mathbb{R}^d$ whereas $\alpha = 1/(d+1)$ for $\mathbb{C}^d$.

2. **Constraints on real equiangular lines.** P. M. Neumann proved that if $\mathbb{R}^d$ contains N equiangular lines with angle $\arccos \alpha$ and $N > 2d$, then $1/\alpha$ is an odd integer [1]. Show that this applies when $d > 3$. Deduce that when $d > 3$, the Gerzon bound can only be attained in $\mathbb{R}^d$ if $d + 2$ is the square of an odd integer.

3. **Equiangular lines in seven Euclidean dimensions.** Consider the following vector in $\mathbb{R}^8$:
$$(3, 3, -1, -1, -1, -1, -1, -1)^{\mathrm{T}}. \tag{7.1}$$

Permute the coordinates of this vector in all possible ways. How many distinct vectors does this yield? Prove that all of them are orthogonal to

© The Author(s), under exclusive license to Springer Nature Switzerland AG 2021
B. C. Stacey, *A First Course in the Sporadic SICs*,
SpringerBriefs in Mathematical Physics 41,
https://doi.org/10.1007/978-3-030-76104-2_7

$$(1, 1, 1, 1, 1, 1, 1, 1)^{\mathrm{T}} \tag{7.2}$$

and thus reside in a 7-dimensional subspace of $\mathbb{R}^8$. What are the angles between them?

4. **SIC optimality** ($\star$). Let $\{R_i\}$ be a minimal informationally complete POVM with Born matrix Φ. Prove that the sum of all the elements of Φ must be d^2, and that Φ must have an eigenvector with corresponding eigenvalue 1. Consider the squared Frobenius distance between I and Φ:

$$||I - \Phi||_2^2 := \mathrm{tr}[(I - \Phi)^\dagger (I - \Phi)] . \tag{7.3}$$

Verify that if $\{R_i\}$ is a SIC, then

$$||I - \Phi_{\mathrm{SIC}}||_2^2 = d^2(d^2 - 1) . \tag{7.4}$$

Prove that in general, the squared Frobenius distance is bounded below:

$$||I - \Phi||_2^2 \geq \frac{(\mathrm{tr}\, \Phi - d^2)^2}{d^2 - 1} . \tag{7.5}$$

The Cauchy–Schwarz inequality may be useful here. To understand the trace of Φ better, note that we can write $\Phi = AG^{-1}$, where A is diagonal and $[G]_{ij} = \mathrm{tr} R_i R_j$. Moreover, we can write $R_i = a_i \rho_i$, where ρ_i has trace 1. Use these facts to prove that the inverses of the eigenvalues of Φ obey the bound

$$\sum_{j>1} \frac{1}{\lambda_j(\Phi)} \leq d - 1 , \tag{7.6}$$

where $\lambda_1 = 1$ as deduced earlier. From this, prove that the trace of the Born matrix is also bounded below:

$$\mathrm{tr}\, \Phi \geq 1 + (d^2 - 1)(d + 1) . \tag{7.7}$$

Combine your deductions to show that

$$||I - \Phi||_2^2 \geq d^2(d^2 - 1) = ||I - \Phi_{\mathrm{SIC}}||_2^2 . \tag{7.8}$$

5. **Measuring a SIC by the Bell-basis method.** Alice is a physicist with a qubit in her laboratory. To the qubit, she ascribes a density matrix ρ. She wishes to perform a SIC measurement upon her qubit, but the only devices in the cabinet are for measurements in orthonormal bases. Fortunately, one of them is for a Bell-basis measurement upon a pair of qubits. That is, Alice can model its outcomes by the projections onto the vectors

$$\frac{1}{\sqrt{2}}\left(|00\rangle \pm |11\rangle\right), \quad \frac{1}{\sqrt{2}}\left(|01\rangle \pm |10\rangle\right). \tag{7.9}$$

Prove that Alice can implement a SIC measurement upon her original qubit by introducing an extra spare qubit to which the ascribes a SIC fiducial state Π_0 and carrying out a Bell-basis measurement upon the pair of them.

6. **A "nonstandard way" to make a qubit SIC.** Suppose that we try to build a qubit SIC starting with the vector

$$|\pi_0\rangle := \begin{pmatrix} 1 \\ 0 \end{pmatrix}. \tag{7.10}$$

Show that the other three vectors in the SIC must have the form

$$|\pi_j\rangle := \frac{1}{\sqrt{3}}\begin{pmatrix} 1 \\ \sqrt{2}e^{i\theta_j} \end{pmatrix} \tag{7.11}$$

up to overall phase factors. Find a matrix of the form

$$M := \frac{1}{\sqrt{3}}\begin{pmatrix} 1 & ae^{i\alpha} \\ \sqrt{2} & be^{i\beta} \end{pmatrix}, \tag{7.12}$$

with a, b, α and β all real, such that repeatedly applying M to $|\pi_0\rangle$ generates $|\pi_1\rangle$, $|\pi_2\rangle$ and $|\pi_3\rangle$ that together make a SIC. What happens if M is applied more times after that?

7. **Constructing sporadic SICs using Hadamard matrices ($\star$).** An order-d complex Hadamard matrix is a $d \times d$ matrix H with the property that each entry has unit magnitude, and $HH^\dagger = dI_d$. Define $V(H, v)$ to be the set of vectors made by splitting H into rows and multiplying one entry in each row at a time by the parameter v. For example, if

$$H_2 := \begin{pmatrix} 1 & i \\ 1 & -i \end{pmatrix}, \tag{7.13}$$

then $V(H_2, v)$ is the set comprising (v, i), $(v, -i)$, $(1, vi)$ and $(1, -vi)$. Prove that there exists a choice of $v \in \mathbb{C}$ that makes this set of (not unit-norm) vectors equiangular. Let $\omega := e^{2\pi i/3}$ and do the same for the 9 vectors obtained from

$$H_3 := \begin{pmatrix} 1 & 1 & 1 \\ 1 & \omega & \omega^2 \\ \omega & 1 & \omega^2 \end{pmatrix} \tag{7.14}$$

and the 64 vectors obtained from

$$H_8 := \begin{pmatrix} 1 & 1 & 1 & 1 & 1 & 1 & 1 & 1 \\ 1 & -1 & 1 & -1 & 1 & -1 & 1 & -1 \\ 1 & 1 & -1 & -1 & 1 & 1 & -1 & -1 \\ 1 & -1 & -1 & 1 & 1 & -1 & -1 & 1 \\ 1 & 1 & 1 & 1 & -1 & -1 & -1 & -1 \\ 1 & -1 & 1 & -1 & -1 & 1 & -1 & 1 \\ 1 & 1 & -1 & -1 & -1 & -1 & 1 & 1 \\ 1 & -1 & -1 & 1 & -1 & 1 & 1 & -1 \end{pmatrix} . \tag{7.15}$$

8. **Manipulations of a qubit SIC.** Consider the qubit state

$$\Pi_0 := \frac{1}{2}\left(I + \frac{1}{\sqrt{3}}(\sigma_x + \sigma_y + \sigma_z) \right) . \tag{7.16}$$

Show that conjugating this state by each of the Pauli operators—that is, $\sigma_x \Pi_0 \sigma_x^\dagger$ and so forth—produces a SIC. Find the vectors $|\pi_j\rangle$ that satisfy

$$\Pi_j = |\pi_j\rangle\langle\pi_j| . \tag{7.17}$$

Let Π_0^* be the state found by taking the complex conjugate of each element in the matrix Π_0. What is Π_0^* in terms of the Pauli operators, and what is its orbit under conjugation by the Pauli group? What are the inner products between elements of one orbit and elements of the other? What shape do the two orbits make together in the Bloch-ball picture?

9. **2-design property.** The pure states for a pair of qubits are rays in $\mathbb{C}^4$. Within this space, we can carve out a *completely symmetric* subspace. If $|0\rangle$ and $|1\rangle$ are the eigenvectors of the Pauli σ_z operator, then their tensor products give a basis for $\mathbb{C}^4$, and the vectors

$$|0\rangle \otimes |0\rangle, \quad \frac{1}{\sqrt{2}}(|0\rangle \otimes |1\rangle + |1\rangle \otimes |0\rangle), \quad |1\rangle \otimes |1\rangle \tag{7.18}$$

are invariant under exchanges of the two qubits. Let $\{\Pi_j\}$ be a qubit SIC. First, verify that

$$\sum_j \Pi_j \otimes \Pi_j = \frac{4}{3} P_{\text{sym}} , \tag{7.19}$$

where P_{sym} is the projector onto the completely symmetric subspace. Then, find what happens when the sum is changed to

$$\sum_j \Pi_j \otimes \Pi_j^{\mathsf{T}} . \tag{7.20}$$

10. **SIC representations of qubit unitaries.** Just as fixing a reference measurement lets us represent quantum states with probability distributions, it also lets us write

unitary operations as matrices of *conditional* probabilities. Let

$$P(R_i) = \frac{1}{d}\text{tr}(\rho\,\Pi_i) \tag{7.21}$$

as usual, and let the measurement $\{D_j\}$ be the act of waiting around a while before using the same device that provides the reference measurement. Then, if the time evolution of ρ is given by $U\rho U^\dagger$,

$$Q(D_j) = (d+1)\sum_{i=1}^{d^2} P(R_i)P(D_j|R_i) - \frac{1}{d}, \tag{7.22}$$

where

$$P(D_j|R_i) := \frac{1}{d}\text{tr}(U\,\Pi_i\,U^\dagger\Pi_j). \tag{7.23}$$

What property of $P(D_j|R_i)$ allows the sum over the second term in the urgleichung to be simplified to $-1/d$? Calculate the conditional probabilities for the Pauli operators and for the Hadamard gate

$$H := \frac{1}{\sqrt{2}}\begin{pmatrix} 1 & 1 \\ 1 & -1 \end{pmatrix}. \tag{7.24}$$

11. **Tensorhedron MIC** ($\star$). Let $\{\Pi_j\}$ be a qubit SIC. Consider a collection of N qubits, and define the operators

$$E_{j_1,\ldots,j_N} := E_{j_1} \otimes \cdots \otimes E_{j_N}, \tag{7.25}$$

where $E_j := \frac{1}{2}\Pi_j$. Prove that the operators $\{E_{j_1,\ldots,j_N}\}$ form a POVM. What is its Gram matrix, and what are the eigenvalues thereof? Show that this POVM is informationally complete. What is its Born matrix Φ, and how does the Frobenius distance $||I - \Phi||_2$ compare to that for Φ_{SIC}?

12. **Wigner functions.** Let $\{\Pi_j\}$ be a SIC in dimension d, and consider the operators

$$Q_j^\pm := \pm\sqrt{d+1}\,\Pi_j + \frac{1 \mp \sqrt{d+1}}{d}I. \tag{7.26}$$

Prove that Q_j^+ and Q_k^+ are orthogonal whenever $j \neq k$, and likewise for Q_j^- and Q_k^-. What are the sums $\sum_j Q_j^+$ and $\sum_j Q_j^-$? Find an expression for

$$W_\rho^\pm(j) := \text{tr}(Q_j^\pm\rho) \tag{7.27}$$

in terms of the SIC probabilities

$$p_\rho(j) = \frac{1}{d}\mathrm{tr}(\Pi_j \rho).\tag{7.28}$$

What is the Euclidean inner product of W_ρ and W_σ for two density matrices ρ and σ?

13. **Qutrit Wigner function.** Let $\{\Pi_j : j = 1, \ldots, 9\}$ be the Hesse SIC, and define

$$W_\rho(j) := \mathrm{tr}(Q_j^+ \rho),\tag{7.29}$$

where Q_j^+ is as defined in the previous problem. Find W_ρ for the Hesse SIC states themselves and for the Norrell states. Recall the construction of four Mutually Unbiased Bases from the Hesse SIC, and let each such basis define a measurement. Relate the probabilities for the outcomes of these measurements to W_ρ. The 3×3 grid may make an appearance.

14. **Norrell states ($\star$).** Consider the vector

$$|v_0\rangle := \frac{1}{\sqrt{2}}\begin{pmatrix} 0 \\ 1 \\ 1 \end{pmatrix}.\tag{7.30}$$

Show that $|v_0\rangle$ is one of the Norrell states. Find its probabilistic representation using the Hesse SIC as reference measurement. Do the same for the other eight states in the Weyl–Heisenberg orbit of $|v_0\rangle$ and verify that they form a SIC. Using the results of the previous problem, calculate the W-functions for the nine states $\{|v_j\rangle\}$. What do they have in common with the W-functions of the Hesse SIC states? How do the states $\{|v_j\rangle\}$ look when their p's and W's are written in the famous 3×3 grid? With that as a guide, what might the p's and W's of the other Norrell states look like? Remember, the Norrell states are a Clifford orbit of four SICs containing nine vectors apiece.

15. **Quantum ellipsoid.** A matrix is positive semidefinite if and only if it can be written as the square of a Hermitian matrix. So, for any quantum state ρ, we must have

$$\rho = B^2\tag{7.31}$$

for some Hermitian B. Define the coefficients $\{b_j : j = 1, \ldots, N\}$ by

$$B = \sum_j b_j \Pi_j,\tag{7.32}$$

where $\{\Pi_j\}$ is a SIC. Show that the condition $\mathrm{tr}\rho = 1$ implies that the vector of the $\{b_j\}$ lies on an ellipsoid. Find an expression for $p(k) := \frac{1}{d}\mathrm{tr}(\rho\Pi_k)$ in terms of the $\{b_j\}$.

16. **The Flammia–Jones–Linden theorem ($\star$).** Let A be a $d \times d$ Hermitian matrix. Prove that A is a rank-1 projection operator if and only if

$$\mathrm{tr}A^2 = \mathrm{tr}A^3 = 1 . \tag{7.33}$$

Having proved the Flammia–Jones–Linden theorem, apply it to the SIC representation of quantum states. Let $\{\Pi_j : j = 1, \ldots, d^2\}$ be a SIC and

$$a(j) := \frac{1}{d}\mathrm{tr}(A\Pi_j), \tag{7.34}$$

so that

$$A = \sum_j \left[(d+1)a(j) - \frac{1}{d} \right] \Pi_j . \tag{7.35}$$

Show that the conditions in the Flammia–Jones–Linden theorem take the form

$$\sum_j a(j)^2 = \frac{2}{d(d+1)} , \tag{7.36}$$

$$\sum_{jkl} [\mathrm{Re}\,\mathrm{tr}\Pi_j \Pi_k \Pi_l] a(j)a(k)a(l) = \frac{d+7}{(d+1)^3} . \tag{7.37}$$

17. **Qubit state space.** Show that in dimension $d = 2$, the QBic Eq. (7.37) is redundant with the quadratic condition (7.36).
18. **Qutrit state space** ($\star$). Let $\{\Pi_j : j = 1, \ldots, 9\}$ be the Hesse SIC. Show that the QBic equation reduces to the expression using the Steiner triple system.
19. **Quasi-SICs** ($\star$). Consider the set of $d \times d$ Hermitian matrices with zero trace. Show that with the Hilbert–Schmidt inner product, this set is a vector space with dimension $d^2 - 1$. In such a space, we can always find a set of points $\{B_j : j = 1, \ldots, d^2\}$ that have norm 1 and form the vertices of a regular simplex. Show that for any two distinct B_j and B_k, their inner product is

$$\mathrm{tr}B_j B_k = -\frac{1}{d^2 - 1} . \tag{7.38}$$

Define

$$Q_j := \sqrt{\frac{d-1}{d}} B_j + \frac{1}{d}I , \tag{7.39}$$

and show that the matrices $\{Q_j\}$ satisfy the inner-product condition of a SIC. Prove that such a *quasi-SIC* $\{Q_j\}$ will *be* a SIC in $d = 2$ but generally won't in higher dimensions.
20. **Deriving the QBic equation by way of quasi-SICs.** Let e_n be a probability vector that has the form of a SIC representation of a SIC state:

$$e_n(j) = \frac{1}{d(d+1)} + \frac{\delta_{nj}}{d+1} . \tag{7.40}$$

Suppose that $\{Q_n\}$ is a quasi-SIC. Show that

$$\sum_{jkl} e_n(j)e_n(k)e_n(l)\operatorname{Re} \operatorname{tr} Q_j Q_k Q_l = \frac{d+6+\operatorname{Re} \operatorname{tr} Q_n^3}{(d+1)^3}. \tag{7.41}$$

21. **P-incompatibility in 2 and 3 dimensions.** Define a qubit SIC $\{\Pi_j^+\}$ using Eq. (7.16). Verify that these four states are S-incompatible with respect to the SIC $\{\Pi_j^-\}$. In $d = 3$, show that any three states from the Hesse SIC are O-incompatible with respect to a measurement in one of the MUB we constructed in the text. What is the difference between triples that are collinear in the infamous 3×3 grid and triples that are not?

22. **Algebraic properties of SIC triple products.** For any SIC, we can define

$$C_{jkl} := \operatorname{Re} \operatorname{tr} \Pi_j \Pi_k \Pi_l . \tag{7.42}$$

Prove that for a group covariant SIC, we can reconstruct all the $\{C_{jkl}\}$ knowing only the $\{C_{0kl}\}$. Define the matrix C_0 by

$$[C_0]_{kl} := C_{0kl}. \tag{7.43}$$

For the qubit, Hesse and Hoggar SICs, what are the eigenvalues of C_0?

23. **Kochen–Specker proof for qutrits.** Verify the bound invoked in the no-hidden-variables proof for qutrits, Eq. (4.45).

24. **Integer lattices.** In the text, we explored the relation between the Hoggar-type SICs and the octavians, which (up to a scale factor) furnish the E_8 lattice. Find the analogous lattices related to the qubit and Hesse SICs, and relate the resulting list of lattices to the classification of Dynkin diagrams outlined in Sect. 6.1.

25. **SICs and number theory** ($\star$). It has been observed empirically that in $d \geq 4$, Weyl–Heisenberg SICs generate number fields that are extensions of $\mathbb{Q}(\sqrt{D})$, where D is the square-free part of $(d - 3)(d + 1)$. This hints at subtle relations between SICs in different dimensions, since different values of d can lead to the same number field $\mathbb{Q}(\sqrt{D})$. Verify that $d = 4, d = 8$ and $d = 19$ imply the same value of D. What number comes next in that sequence? A more sophisticated analysis finds that for an arbitrary square-free integer $D \geq 2$, we can always find a dimension $d \geq 4$ such that the square-free part of $(d - 3)(d + 1)$ is equal to D. If d_1 is the smallest such integer, then we have in fact a whole sequence of integers d_j that satisfy the same property. These are given by

$$d_j = 1 + 2T_j \left(\frac{d_1 - 1}{2}\right), \tag{7.44}$$

where T_j is the jth Chebyshev polynomial of the first kind. Using the recursion relation for Chebyshev polynomials,

$$T_{j+1}(x) = 2x T_j(x) - T_{j-1}(x),\qquad(7.45)$$

deduce a recursion relation for d_{j+1} in terms of d_1, d_{j-1} and d_j. The result you find should be *inhomogeneous,* having a constant term. Show that you can eliminate this constant offset by converting to a homogeneous recursion relation of higher order, yielding d_{j+1} in terms of d_j, d_{j-1} and d_{j-2}. Find the limit of the ratio d_{j+1}/d_j as j tends to infinity. (A text on continued fractions may be helpful.) What is this limit when $d_1 = 4$?

26. **Open-ended exploration of the literature ($\star\star$).** Start with Zauner's thesis [2], the 2004 paper that introduced SICs to the physics community [3], and Scott and Grassl's study [4]. Using the database of your choice, see what writings cite these, and map out the subject area. In what journals have people published on SICs? What papers have been most influential? (What should one mean by "influential"?) Do the citation trails lead anywhere that surprises you?

References

1. P.W.H. Lemmens, J.J. Seidel, Equiangular lines. J. Algebra **24**(3), 494–512 (1973)
2. G. Zauner, Quantendesigns. Grundzüge einer nichtkommutativen Designtheorie. Ph.D. thesis, University of Vienna (1999). https://doi.org/10.1142/S0219749911006776. http://www.gerhardzauner.at/qdmye.html; Published in English translation: G. Zauner, Quantum designs: foundations of a noncommutative design theory. Int. J. Quantum Inf. **9**, 445–508 (2011)
3. J.M. Renes, R. Blume-Kohout, A.J. Scott, C.M. Caves, Symmetric informationally complete quantum measurements. J. Math. Phys. **45**, 2171–2180 (2004). https://doi.org/10.1063/1.1737053
4. A.J. Scott, M. Grassl, Symmetric informationally complete positive-operator-valued measures: A new computer study. J. Math. Phys. **51**, 042203 (2010). https://doi.org/10.1063/1.3374022

Index

A
Albert algebra, 95
Algebraic number theory, 5, 8, 94, 110

B
Balanced incomplete block design, 61, 70
Bell inequality, 7, 14, 39–41, 43, 44, 50, 52
Biodiversity, 20, 95
Bitangents, 3, 78, 90, 91
Bloch ball, 22, 27, 28, 31, 32, 49, 50, 88, 90, 95, 106
Bloch sphere, 45
Born matrix, 18, 104, 107
Born rule, 6, 15, 16, 18, 33, 51

C
Cauchy–Schwarz inequality, 21, 67, 94, 104
Chebyshev polynomials, 110
Clebsch surface, 87
Clifford group, 23, 35, 60, 72, 108
Combinatorial design, 61, 69, 75
Completely symmetric subspace, 106
Conway groups, 5, 94
Cubic surface, 87
Cyclotomic fields, 5, 9

D
Dedekind lattice, 96
Double-slit experiment, 13, 36, 44, 52
Doubly transitive, 28, 35, 62, 94
Dual frame, 17

Dynkin diagram, 84, 87, 88, 91, 110

E
Effective number, 20
Eisenstein integers, 58
Entanglement, 39
Entanglement-breaking channel, 19
EPR criterion, 45, 46
Exceptional Lie algebras, 83
Exclusive or, 75, 91

F
Fano plane, 3, 58, 79, 91, 97, 99
Fibonacci–Lucas SICs, 94, 111
Flammia–Jones–Linden theorem, 32, 108
Frame theory, 16, 17
Frobenius distance, 19, 104

G
Galois group, 8, 87
Generalized polygons, 99
Gerzon bound, 2, 91, 92, 94, 95, 103
GHZ state, 40, 43, 44
Gleason's theorem, 51
Golden field, 87, 94
Golden ratio, 2, 93, 111
Gossett polytope, 3, 104
Gröbner bases, 8
Group covariance, 7, 22, 28, 34, 36, 41, 60

© The Author(s), under exclusive license to Springer Nature Switzerland AG 2021   113
B. C. Stacey, *A First Course in the Sporadic SICs*,
SpringerBriefs in Mathematical Physics 41,
https://doi.org/10.1007/978-3-030-76104-2

H

Hadamard gate, 107
Hadamard matrix, 70, 105
Hall–Janko group, 98
Hesse configuration, 86
Hesse SIC, 21, 23, 24, 28–30, 34, 50, 59, 72, 88, 105, 108–110
Hessian polyhedron, 86, 89
Hidden variables, 7, 36, 40, 43, 44, 46, 47, 49, 75
Higman–Sims group, 5, 95
Hilbert's twelfth problem, 9
Hoggar-type SICs, 9, 21, 24, 29, 34–36, 41, 42, 52, 59–61, 67, 69, 75, 89, 105, 110
Hurwitz integers, 58
Hurwitz's theorem, 58

I

Incompatible quantum states, 72, 110
Informational completeness, 7, 16, 29

K

Kochen–Specker theorem, 7, 34, 50, 80, 110
Kronecker–Weber theorem, 9

L

Law of Total Probability, 15, 17–19
Leech lattice, 3, 4, 7, 94, 95
Lie algebra, 17, 85, 91
Lie group, 85

M

Mathieu groups, 5
McLaughlin group, 5
Minimum uncertainty states, 33, 35
Monster group, 99
MUB-balanced states, 33, 35
Mutually Unbiased Bases, 29, 32, 34, 50, 73, 75, 88, 95, 108

N

Norrell states, 24, 108

O

Octavians, 8, 58, 97, 110
Octonions, 3, 8, 58, 91, 95, 97

P

Pauli group, 8, 27, 31, 34–36, 40, 41, 43, 61, 67, 78, 89, 90, 99, 106, 107
Post-selection, 44
POVM, 6, 15, 17, 28, 45, 48, 72

Q

QBic equation, 20, 22, 28, 29, 31, 60, 109
Quantum coherence, 31, 36
Quantum computation, 34, 50
Quantum interference, 13, 44
Quantum logic, 96
Quantum random number generation, 50
Quasi-probability, 19, 49
Quasi-SIC, 109, 110
Quaternions, 58
Qubit, 27, 31, 40, 45, 88, 104–107, 109, 110
Qubit SIC, 9, 21, 22, 27, 29, 32, 48–50, 59, 70, 88, 90, 105–107, 109, 110
Qutrit, 23, 28, 34, 40, 50, 95, 108–110

R

Ray class fields, 9, 94
Rebit, 6, 94, 96
Reconstructing quantum theory, 6, 39, 52
Reference measurement, 14–16
Regular hexagon, 1
Regular icosahedron, 2, 93
Regular tetrahedron, 22, 27, 28, 32, 70, 88, 92
Relative entropy of coherence, 33
Rényi entropy, 36
Resource theory, 20, 31, 34, 36, 50
Reuglar tetrahedron, 45
Root system, 83, 89
Rudvalis group, 99

S

Seidel adjacency matrix, 67
Shannon entropy, 21, 27–29, 67, 88, 90
SICs in the lab, 17
Singularity, 88
Special relativity, 53
Spekkens toy model, 36, 44, 52, 96
Sphere packing, 4, 8, 88, 90
Stabilizer group, 5, 35, 59
Standard Model, 96
Stark units, 9
Steiner triple system, 29, 109
Strongly regular graphs, 5
Switching class, 5

Symmetric design, 62
Symmetric difference, 71, 76
Symplectic design, 70, 71, 77

T
Tetrahedral group, 92
Thermodynamics, 53
Trine, 88
Triple products, 20, 28, 30, 59, 60, 62, 64, 65, 77, 110
2-design, 106
Two-graph, 66, 67

U
Unitarily invariant norm, 19
Urgleichung, 17–19, 43, 45, 52, 53, 107

V
von Neumann entropy, 34, 35

W
Weyl group, 84, 87, 91
Weyl–Heisenberg group, 8, 22, 60, 92
Wigner functions, 17, 20, 107, 108
Witting polytope, 89

Printed in the United States
by Baker & Taylor Publisher Services